ELECTROTECHNOLOGY VOLUME 8

BIOTECHNOLOGY AND ENERGY USE

ELECTROTECHNOLOGY VOLUME 8

BIOTECHNOLOGY AND ENERGY USE

ROBERT J. CLERMAN
RAJANI JOGLEKAR
ROBERT P. OUELLETTE
METREK Division of the MITRE Corporation
McLean, Virginia

PAUL N. CHEREMISINOFF
New Jersey Institute of Technology
Newark, New Jersey

ANN ARBOR SCIENCE
PUBLISHERS INC / THE BUTTERWORTH GROUP

CHEMISTRY

6654-0641

PREFACE

The objective of this volume is to evaluate the potential impact of bio-industry on industrial energy use. In this survey more than 60 biotechnology applications in key industrial sectors (e.g., food, energy and waste treatment) are reviewed. These applications involve advanced uses of fermentation and enzyme technology. This volume continues the series of studies initiated and sponsored by Electricité de France (EdF), in this case, to study the growth of biotechnology applications in industry and the implications for industrial energy patterns.

There is little doubt that a "revolution" in applied biology is now under-way. The most publicized and far-reaching of the developments fostering this revolution is recombinant DNA technology, commonly referred to by the general term "genetic engineering." This technology, which involves transfer of genetic material between biologically divergent organisms, allows scientists to alter cells to perform new functions. The implications of this breakthrough are just starting to be realized in a wide range of industries, including manufacturers in the specialty and commodity chemicals, phar-maceuticals, food processing, energy-related and agricultural industries.

The dramatic increase in the commercial potential of biological processes is not, however, limited to breakthroughs in genetic engineering. More rele-vant for the near-term are recent developments in biotechnology related to fermentation and the industrial use of enzymes. For example, the global market for industrial enzymes is expected to grow at an annual rate of 8% to a total of approximately $500 million by 1985. Although enzymes have been used in industry for centuries, the projected increase in use can be attributed to biotechnology advances in both production and preparation (e.g., microbial production and development of immobilized enzymes).

One of the most compelling reasons for the broad application of biotech-nology in industry is the opportunity to convert processes that consume fossil fuels (either as fuel or raw materials) to new processes that are energy-efficient or that consume renewable/abundant raw materials.

Our objectives in this study are twofold:

v

1. to identify and describe the range of industrial applications of biotechnologies identified to date; and
2. to identify those applications that have a potential for near-term commercialization and that are likely to affect energy consumption patterns.

What follows in this book is a discussion of the principal tools of biotechnology, the key industrial sectors surveyed, conclusions and brief technical descriptions of individual applications.

For the purposes of this study, bioindustry is a general term, representing a wide range of industrial applications of biological systems (biotechnologies). The emphasis is on the use of microorganisms (whole cells, cellular materials or enzymes) in the manufacture of products or the improvement of processes. At the request of EdF, we have focused our survey on five key industrial sectors: food, energy, waste treatment, chemicals and metals recovery.

It became apparent at an early stage in the project that biotechnology research and development is extensive in the above areas as well as in pharmaceuticals, agriculture and analytical instrumentation. Focusing on these key sectors of interest, we have identified and described applications based on the most recent information available.

We have not included applications based directly on recombinant DNA technology or other aspects of genetic engineering. There are two reasons for this: (1) most proposed commercial applications of genetic engineering are still in the early stages of research and development; and (2) details regarding these processes are considered proprietary business information and therefore are not publicly available.

Computer literature searching is an invaluable tool in surveying a field as diverse and contemporary as bioindustry. The search strategy basically involved matching descriptors relating to food, energy, waste treatment, chemicals and metals recovery with terms relating to microbial processes. A special descriptor, "bioconversion" or "biotechnology," increased the search capability, provided such a descriptor was included in the research citation. The computer literature search is limited, however, in that: (1) it dates back only six to seven years, (2) only widely distributed literature is generally covered, and (3) indexing is sometimes not sufficient to provide comprehensive coverage. We found that, although time-consuming, manual searching and cross-referencing is essential to supplement the computer literature searches and achieve adequate coverage in this field.

Following the literature search, more than 200 citations were scrutinized, and articles or monographs were obtained. In reviewing this information, individual microbial processes were selected for summary if they possessed one or more of the following characteristics either directly or in comparison with a conventional counterpart:

- potential savings in time or materials requirements,
- means for conversion from batch to continuous processing,

- economically significant improvement in product quality,
- capability for production of more than one product simultaneously or serially in the same process,
- commercial feasibility, and
- potential reduction (or increase) in energy consumption.

In most cases, the selected applications meet more than one criterion.

In all cases the objective was to obtain: (1) a basic description of the process (including flow diagrams wherever possible), (2) current stage of development (laboratory, pilot, demonstration, or commercial), and (3) an initial assessment of potential implications for energy consumption. In many cases, the energy implications are stated as "unclear at present." This does not necessarily imply that energy impacts will be nonexistent, but rather that a more detailed analysis is required before a judgment can be made. When questions arose regarding specific aspects of a process, its progress toward commercialization, or its energy implications, researchers or industry representatives were contacted directly.

The authors gratefully acknowledge Electricité de France for support in this work.

Robert J. Clerman
Rajani Joglekar
Robert P. Ouellette
Paul N. Cheremisinoff

| Clerman | Joglekar | Ouellette | Cheremisinoff |

Robert J. Clerman is a group leader in the Environment Division of the MITRE Corporation. He received his bachelor's degree in biology from the State University of New York at Fredonia, and has a master's degree in environmental sciences from the University of Virginia. His research interests and publications have covered a wide range of environmental issues. Since joining MITRE, Mr. Clerman has been involved in studies relating to the fate and effects of chemicals in the environment. Projects have included development of a tiered testing system for new chemicals evaluation, design of a residuals monitoring program (both for the Federal Republic of Germany), and evaluation of organic pollutants in water (for the Department of Energy and the National Cancer Institute). Currently he is managing studies of industrial applications of biotechnology.

Rajani Joglekar received her BSc (biology) from the University of Bombay, a MS (biology) and MS (environmental sciences) from Northeastern University and George Washington University, respectively. Her research interests and publications are in the areas of plant growth and development, and development of biological tests for detecting toxic chemicals in the environment. Since 1978, working as a member of the technical staff in the Environment Division of the MITRE Corp., she has been studying the health and environmental effects of energy technologies. These studies have included the health effects of coal and oil shale, health and environmental effects of synthetic fuels, oil and gas-end use, and development of research plans for coal liquefaction technologies. Currently, Ms. Joglekar is involved in evaluating the industrial applications of biotechnology.

Robert P. Ouellette is Technical Director, Environment Division, of the MITRE Corp. Dr. Ouellette has been associated with MITRE in varying capacities since 1969. Earlier, he was with TRW Systems, Hazelton Labs Inc. and Massachusetts General Hospital. A graduate of the University of Montreal, he received his PhD from the University of Ottawa. A member of the American Statistical Association, Biometrics Society, Atomic Industrial

Forum and the National Science Foundation Technical Advisory Panel on Hazardous Substances, Dr. Ouellette has published numerous technical papers and books on energy and the environment and is co-editor, co-author of the Electrotechnology series published by Ann Arbor Science.

Paul N. Cheremisinoff is Associate Professor of Environmental Engineering at New Jersey Institute of Technology. He is a consulting engineer and has been consultant on environmental/energy/resources projects for the MITRE Corp. He has more than 30 years of practical design, research/development and engineering experience in a wide range of industries including pollution control, waste treatment, chemical and process industries. He is author/ editor of many Ann Arbor Science publications, including: *Pollution Engineering Practice Handbook*, *Carbon Adsorption Handbook* and *Environmental Impact Data Book*.

CONTENTS

LIST OF FIGURES

LIST OF TABLES

CHAPTER 1

OVERVIEW OF BIOTECHNOLOGY APPLICATIONS

PRINCIPAL TOOLS OF BIOTECHNOLOGY

Bioindustry can be broken down into several component technologies that apply across a wide range of applications. Whole cell fermentation processes harness the power of microorganisms to chemically transform substances. Enzyme technology involves the isolation and preparation of microbial enzymes to catalyze important reactions. Genetic engineering, the collection of techniques for transfer of genetic material between disparate organisms, could affect virtually every aspect of bioindustry by allowing researchers to manipulate organisms to perform completely new functions.

These three tools of biotechnology—whole cell fermentation, enzyme technology and genetic engineering—are discussed briefly, followed by an overview of the major industrial sectors in which they have been applied.

Whole Cell Fermentation

Microbial fermentation processes have been commonplace for centuries. Wine-making and baking were refined arts (and important industries) long before the advent of microbiology as a science. Although the range of applications has increased and the techniques have been refined, until recently the fermentation industry has not been characterized by the rapid growth and innovation typical of industries such as synthetic chemicals or electronics. This situation appears to be changing, and current growth and innovation in whole cell fermentation may be attributable to two recent phenomena: (1) environmental and economic incentives to use organic wastes in favor of petroleum-based feedstocks, and (2) advances in microbiology such as genetic engineering.

The details of recent developments in the industrial application of whole cell fermentation are discussed later. We have found active commercial research and development in areas such as:

- improvement of food quality
- production of sweeteners and protein supplements
- transformation of organic wastes into fuels
- waste treatment
- metals extraction

Enzyme Technology

From an industrial viewpoint, microbes are organized systems of enzymes that catalyze key chemical reactions. These enzymes, separated from the microbial cells, represent an important tool of biotechnology. Just as with whole cell fermentation, enzyme technology is undergoing a period of rapid advancement and commercial growth potential. This growth is due principally to innovations in the preparation and use of immobilized enzymes.

Enzyme immobilization is a novel technique in which enzymes are adsorbed, copolymerized, covalently attached or otherwise associated with an inert or water-insoluble material. This allows the enzyme to be separated physically from both the substrate and product and to be recycled or reused following cleansing. The major advantages of immobilized enzyme systems include the following [1]:

- multiple use of a single batch of enzyme,
- improved process control,
- enhanced stability, and
- freedom from enzyme contamination of products.

Immobilized enzymes of microbial origin have been used successfully in food processing, fermentation processes, pharmaceuticals production and development of analytical instruments. In this study, commercially viable applications of immobilized enzymes were found to be most prevalent in the dairy and sweeteners sectors of the food processing industry. Perhaps the most successful application to date has been the use of glucose isomerase enzyme systems in the production of high-fructose corn syrup.

The return on investment from any given application of immobilized enzymes must be significant to justify the investment required for development and optimization of an enzyme system. Before immobilized enzymes can realize their total potential market, improvements need to be made in the techniques of immobilization. This is currently an area of active research and development.

Genetic Engineering

As with fermentation and enzyme technology, genetic engineering is a basic biological tool which can be applied to a range of industrial processes. However, the impact of recent developments in genetic engineering is likely to exceed that of any other development in biotechnology. Through the use of recombinant DNA techniques, researchers can transfer segments of DNA containing desired genetic information from one species to another to create a population of altered, identical cells that perform completely new functions. The potential now exists to improve virtually any biological process (including those reviewed in this study) and to create totally new applications for biological systems in industry.

In Japan, the United States and Europe, recombinant DNA technology is moving rapidly from a research tool in the laboratory into a commercial development phase. The applications are in industries as diverse as foods, energy, chemicals, pharmaceuticals and mining. Bridging the gap between basic research and commercial application is a group of small, venture-capital firms that have emerged over the past decade. These research corporations, including Cetus, Genentech, Genex and Biogen, in many cases, are involved in joint ventures with larger corporations such as Hoffman La Roche, Standard Oil and Eli Lilly. The current status of commercial development in genetic engineering might be typified by the projects underway at Cetus Corporation, the largest of the genetic engineering firms.

Cetus is involved in three major commercial development ventures in conjunction with major U.S. corporations:

1. The first is a novel process to convert cornstarch directly to crystalline fructose and ultrapure fructose syrups, both of high value to a food industry with a market estimated at $7 billion annually.
2. A process to oxygenate certain petrochemical feedstocks, enabling, for example, the conversion of ethylene to ethylene oxide or ethylene glycol, is second.
3. Third is a low-cost, high-yield method for the production of human interferon.

Besides these specific projects, Cetus is also involved in other projects such as continuous fermentation of ethanol for fuel and chemical feedstocks, production of vaccines against malaria and other parasitic diseases, and biomass conversion processes for chemicals and energy applications [2]. Although research and development in these and other areas is progressing rapidly, it is too early to determine with accuracy the extent to which commercial ventures will be successful over the near term.

KEY INDUSTRIAL SECTORS STUDIED

The majority of bioindustry applications surveyed were in three major industrial sectors: (1) food processing and production, (2) energy production, and (3) waste treatment. In addition, recent developments in chemicals, metals recovery and analytical instrumentation were surveyed. In the sections that follow, the status of biotechnology in these industrial sectors is discussed.

Foods

Fermentation technology has been used extensively in the production of dairy and baked goods and traditional alcoholic beverages. In the last 20 years, researchers have identified various biochemical pathways in the fermentation process and improved understanding of the role of microbial enzymes and their mode of action under different operating conditions. This knowledge has been applied extensively in the dairy industry, where the efficiencies of conventional microbial processes are being improved by:

- improved culture preparation,
- use of thermophilic microorganisms,
- use of immobilized enzymes, and
- conversion from batch to continuous production.

(Dairy and other food industry applications are included in Chapter 2.)

Conventional methods for production of sweeteners are typically characterized by high production and raw materials costs. Advances in this sector of the food processing industry are typified by the use of immobilized glucose isomerase enzyme in the production of high-fructose corn syrup from cornstarch. This process has allowed manufacturers to replace conventional soluble enzyme systems with a continuous, automatically controlled system. The reduction in price of fructose syrup has been enough to make it a viable substitute for sucrose derived from sugarcane or sugarbeets.

Production of single-cell protein (SCP) refers to the culture of yeasts, bacteria, fungi and algae for their protein content. In recent years, SCP has been viewed as a supplemental protein source, either for animal feed or for direct human consumption. Although the technology is well advanced, there are a number of factors mitigating against commercial acceptance of SCP, especially for human consumption. These include:

- high capital investment required,
- energy consumption costs to maintain sterile conditions,
- high nucleic acids content of SCP products, which can cause health problems,
- requirement for abundant sources of raw materials to achieve economies of scale, and
- lack of public or regulatory acceptance of SCP products, especially those derived from bacteria.

Some of these problems are now being overcome through advances in process design and a shift to use of abundant, low-cost wastes as substrate.

The trend toward use of waste materials in SCP production is indicated by the fact that 9 out of 14 applications surveyed use agricultural, urban or industrial waste materials. Typical of the industry as it stands today is the application entitled: "Waste Biomass as a Source of Single-Cell Protein (Waterloo Process)" (see Chapter 2). The process uses cellulosic wastes such as agricultural, animal and forestry residues to culture a fungus that has a nucleic acid content that meets standards for human consumption. Low operating temperatures and pressures and an option for recycling methane gas from animal wastes make the commercial-scale version of this process energy efficient.

The two additional applications identified in the food industry are production of wine and edible oils from cheese whey. These two processes illustrate the value of biotechnology in conversion of a waste material, which otherwise would cause disposal problems, into a valuable end product. The wine production system is particularly attractive to the cheese industry since it offers a whey by-product without the energy-intensive process of water removal.

Energy

Biotechnology is playing an increasingly important role in the field of energy development. Applications under development reflect renewed interest in conventional sources such as methane and ethanol, accelerated research into new sources (e.g., hydrogen from algae), and technical breakthroughs in recovery of oil reserves. As in other industrial sectors surveyed, biotechnology applications in energy are characterized by an emphasis on the use of low-cost materials, particularly waste products.

Production of methane gas by the microbial conversion of organic materials to methane is a natural process, providing fuel in a clean, gaseous form. The production of methane from the anaerobic fermentation of organic residues such as sewage sludge, cow manure, food processing waste and urban refuse is in common practice. The process typically involves enzymatic hydrolysis of these complex substances to form soluble organic compounds, followed by an acid hydrolysis catalyzed by bacteria. Finally, the resultant organic acids undergo decomposition as a result of methanogenic bacterial action to produce methane. Depending on the organic content of the refuse, 10–70% of the energy-generating organics can be converted to methane gas. The technology of methane (biogas) production is highly developed in parts of Asia and in Europe. Additional research in reactor designs would reduce the costs of this technology and make it more accessible to small-scale operators. In particular, advancements are being made in accelerating the rate of conversion to methane and reducing temperature maintenance requirements.

Production of ethanol for fuel is another example of a basic biological process undergoing extensive refinement and commercial development. The applications described in Table 1-1 and Chapter 3 focus primarily on the conversion of cellulosic wastes to sugars, which then can be fermented to produce ethanol.

A variety of microorganisms produce hydrogen. One of the emerging areas of biotechnology is the development of a commercially feasible method of harnessing this capability. Additionally, in theory, algae and photosynthetic bacteria can hydrogenate organic substrates to produce gas and oil in the presence of sunlight or elevated temperatures. However, there are significant technical and economic hurdles to overcome before these processes can be commercialized.

Biotechnology has been extended to the extraction of fossil energy resources. The use of microbial leaching techniques to extract oil from shale potentially could reduce the environmental and energy consumption problems associated with conventional retorting; however, the process is at the laboratory stage of development and faces technical problems. Polysaccharides produced by microbial cells apparently have a commercial future in the oil industry as flooding fluids in tertiary oil recovery. Examples of all of the above applications can be found in Chapter 3.

Waste Treatment

The increased sophistication of our society over the past several decades has led to dramatic increases in the volume and complexity of wastes. The use of biotechnology in the treatment of these wastes dates back at least 65 years to the development of the activated sludge process. Once limited to the treatment of conventional municipal and industrial wastes, microbial systems are now being used for oil and grease removal, denitrification, pretreatment of industrial wastes and mining site restoration (see Chapter 4). The development of mutant strains and specially designed reactors to handle these waste treatment problems has resulted in a significant new industry [3].

Chemicals

The chemical industry is large and highly diversified, encompassing a wide variety of products and processes. The use of enzyme preparations and fermentation techniques is well established in a number of application areas. As with other industrial sectors, the trend in research and development is toward the use of organic waste materials as feedstock. (Six out of seven applications in Chapter 5 are based on the use of cellulose, whey and other waste materials.)

We have limited the scope of our review of biotechnology in the chemicals

Table 1-1. Some Typical Applications

Industrial Sector	Application	Stage of Development	Description
Food Industry	Production of edible oils from cheese whey	Laboratory	Conversion of carbohydrates in dairy waste (whey) to edible oils by yeast
	Production of sweeteners from cheese whey	Semiindustrial	Use of immobilized enzymes to catalyze conversion of lactose in whey to sweetener
	Waste biomass as a source of single-cell protein	Pilot	Use of agricultural, forestry, and animal wastes to produce fungal proteins for human consumption
Energy	Cellulosic waste treatment to produce alcohol	Prepilot/commercial	Hydrolysis of cellulosic waste to produce sugar and, subsequently, ethanol
	Methane production utilizing cow manure (digestor operation)	Full-scale production	Anaerobic digestion of cow manure
	Production of hydrocarbons (liquid) from algae	Laboratory	Use of green algae to produce gas and oil
Waste Treatment	Use of denitrifying bacteria in municipal waste water treatment	Pilot	Biological fluidized-bed process with greater capacity and efficiency than typical chemical treatment processes
Chemical Industry	Production of ethanol from starch	Commercial	Use of selected enzymes to produce ethanol from starch
Metals Recovery	Uranium extraction by bacterial leaching	Commercial	Use of bacterial percolation to result in uranium solubilization

industry so that we could focus our efforts on foods, energy and waste treatment. There are indications, however, that developments in immobilized enzymes and genetic engineering may have a significant impact on an organic chemicals market measured in terms of billions of dollars.

Metals Recovery

Bacterial leaching is an important, emerging technique for the extraction of metals from their ores. A solubilization principle is the backbone of this methodology. Iron-oxidizing bacteria *(Thiobacillus, Sulfolobus)* function in two ways: ferric ion generation and metal leaching. This technique has been used to extract uranium and copper from low-grade metal ores.

Another application of this technique currently under development is the use of bacteria under aerobic or anaerobic conditions to digest various trace metals from landfill leachates, solid wastes and wastewaters. The methane gas produced under anaerobic digestion can be trapped to use as an energy source. Applications in metals recovery are found in Chapter 6.

Instrumentation

This category includes innovative ways of using microorganisms in biomedical and clinical instrumentation, as well as in analytical instruments. Bacterial electrodes prepared using immobilized microbial enzymes or whole cells can substitute for widely used synthetic enzyme indicators. This particular application may provide reusable, reliable, sensitive, stable and relatively cheap diagnostic and monitoring tools. Instrumentation applications are described in Chapter 7.

STATUS OF BIOINDUSTRY

It is apparent from this initial survey that biotechnology is a highly dynamic field. Growth in the use of microbial systems across a wide range of industries can be attributed to the following general properties of such systems:

1. Microorganisms can function as small chemical factories synthesizing substances, in many cases more economically than by other means.
2. They can often utilize wastes or other low-cost materials as substrates.
3. Immobilized enzymes increase the efficiency of some catalytic reactions.
4. Use of microbial systems can reduce energy consumption due to lower operating temperatures and pressures.

There are, however, some technical hurdles to be overcome as biotechnology is developed commercially. These include:

- difficulty in conversion from batch to continuous processing,
- sensitivity of biological systems to variations in temperature, pH, nutrients, etc.,
- susceptibility to sudden shutdown of some processes due to viral infection or contamination with toxins, and
- quality and quantity control.

Generally, the field of bioindustry is characterized by active research and development at the laboratory and pilot-plant stage of development (Figure 1-1). Our study has indicated, however, that the rate of commercialization is rapid and that in some industries, such as food processing and waste

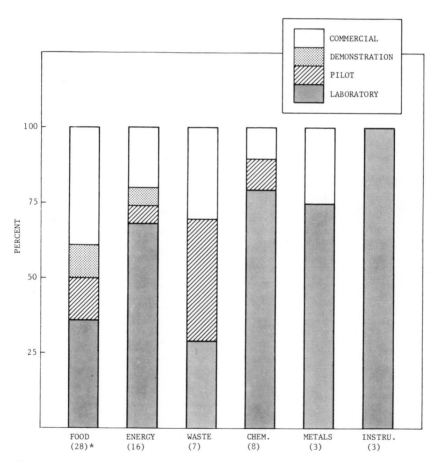

*Number of Projects Surveyed

Figure 1-1.   Stage of development of biotechnology applications surveyed.

treatment, advanced biotechnologies are currently being applied commercially.

THE FUTURE

Based on their commercial potential and/or potential impact on energy consumption, the following industries offer opportunities for biotechnology development:

1. Food Industry
   - Edible oils from cheese whey
   - Wine production from cheese whey waste
   - Sweetener (fructose syrup) production from cornstarch (commercial)
   - Waste biomass as a source of single-cell protein
   - Single-cell protein from cellulosic waste
2. Energy
   - Improved ethanol production through increased fermenter productivity
   - Methane production using dairy manure (digestor operation)
   - Cellulosic waste treatment to produce alcohol
   - Production of hydrocarbon from algae
   - Utilizing urban waste for production of methane gas.
3. Waste Treatment
   - Mutant bacteria in treatment of oily wastes
   - Denitrifying bacteria in municipal wastewater treatment
   - Improved oxygen utilization in wastewater treatment by the aeration tower technology
4. Chemicals
   - Production of ethanol from starch
   - Simultaneous production of ethanol and single-cell protein from cellulosic waste.

CHAPTER 2

FOOD INDUSTRY

A key industrial sector is the food industry. Applications involve advanced uses of fermentation and enzyme technology. This chapter includes brief technical descriptions of individual applications in this key industry.

EDIBLE OILS DERIVED FROM CHEESE WHEY

Description

Until recently, the conversion of carbohydrates to fats and oils by fermentation has not been considered seriously as a commercial process, but the problems of disposal of food processing wastes and by-products make this an attractive option. The fermentation of whey and permeate to oil has been achieved successfully under laboratory conditions by using the yeasts *Candida curvata* and *Trichosporon cutaneum* as catalysts.

The lactose fermentation is carried out in a 10-liter fermenter equipped with temperature, aeration, stirring and pH controls. The nutritional and physical conditions can be optimized for yeast growth and oil production by directly monitoring microscopic counts and respiration rates. The substrate, whole whey, is reconstituted to 6.5% solids and condensed permeate. Approximately 5×10^8 cell/ml are used in the process. The fermentation is carried out for 72 hours, following which lipid extraction is undertaken. Under optimized conditions, it is evident that whey permeate gives greater fermentation yields and efficiencies compared to whole whey, probably due to a greater concentration of lactose in the permeate [4]. It has been suggested that all yeast strains ferment lactose in two phases consisting of a growth phase with little fat accumulation followed by a fattening phase. This

two-stage fermentation results from depletion of growth nutrients and conversion of residual carbohydrates to oil. Figures 2-1 and 2-2 present production of oil by yeast when grown in whole whey and in ultrafiltered whey,
respectively.

Lipid extraction from yeast is difficult, probably due to characteristics of
the cell walls. Methanol: benzene (1:1, v/v) extraction has been shown to be
the most efficient extraction process, with an effective oil yield of 51.1% fat
dry weight. The disadvantages of using methanol and benzene are that they
are expensive and toxic solvents to consider for extraction of edible oils. An
ethanol-hexane mixture may be a more acceptable alternative.

The fatty acid composition of the cells varies with growth temperature,
fermentation time and medium composition. The yeast strains contain
considerable amounts of oleic, palmitic and stearic acids. Oil composition
changes significantly when the permeate is used as a substrate.

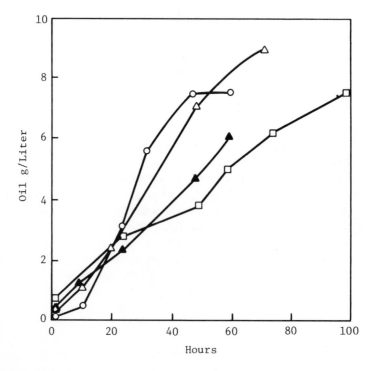

O C. curvata R,   Δ C. curvata D,
▲ T. cutaneum 24,   □ T. cutane

Figure 2-1.   Production of oil by yeasts when grown in whole whey [4].

To make this process attractive for commercialization, more research is needed to improve the efficiency of fermentation. This process of yeast-oil extraction seems to be unattractive to oil processors in the United States. For example, if 1.3×10^7 metric tons of sweet cheese whey were to be converted by this process, 2×10^5 metric tons of oil will be produced. This quantity is only a fraction (5.4%) of the estimated edible oil produced in the United States in 1977 [5]. However, for other countries this amount may be equivalent to the total edible oil requirement.

Stage of Development

This process is at the stage of laboratory-scale research and development.

Implications for Energy Consumption

Agitation and aeration require 50% of the energy needed for the process. Further work in reducing agitation and aeration requirements will result in reduced energy costs.

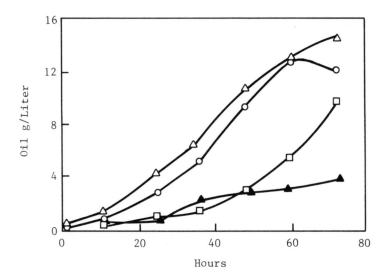

o C. curvata R,   △ C. curvata D,

▲ T. cutaneum 24,   □ cutaneum 40.

Figure 2-2.   Production of oil by yeast when grown in ultrafiltered whey [4].

PRODUCTION OF SWEETENERS FROM CHEESE WHEY (SEMIINDUSTRIAL)

Description

Cheese whey is a major by-product of cheese manufacturing. Today, most whey is spray-dried or concentrated and used in a variety of agricultural or human applications. Cheese whey proteins are presently recovered by ultra-filtration which, in turn, yields a secondary by-product: whey permeate (the liquid fraction after filtration). This permeate fraction is a potential pollution hazard, and disposal creates a problem. Cheese whey permeate does, however, contain 40 g/l of lactose, and a process of hydrolysis has been developed to convert this lactose into an edible sweetener syrup. Hydrolyzed lactose is also useful as a fermentation medium.

Lactose hydrolysis is catalyzed by a α-galactosidase or lactase enzyme, and a technique has been developed to immobilize this lactase enzyme. Immobilized lactase enzyme permits lactose hydrolysis to run on a continuous basis at lower costs than with nonimmobilized enzyme [6]. Immobilized lactase enzyme is now commercially available from Corning and has been in use in a semiindustrial operation [6]. Figure 2-3 indicates possible pathways of lactose hydrolysis in whey upgrading.

The α-galactosidase is extracted from *Aspergillus niger* grown on beet pulp-enriched media. The enzyme is covalently bound to a controlled-pore silica carrier. The immobilized lactose is introduced in a vertical cylindrical column shown in Figure 2-4 and is operated downflow. The process is run at constant throughput and temperature is raised slightly to compensate for the thermal deactivation of the composite. Cleaning is achieved by backflushing the bed with dilute acetic acid [7].

The long-term stability and the economics of the process have been tested under actual industrial operation in a semiindustrial plant since 1978 at the MMB Technical Division facility in Crudgington, England. The plant operates on a continuous basis five days a week and processes about 30,000 liters of cheddar sweet whey a week, resulting in about 1.7 tons of syrup (1200 kg of solids). Figure 2-5 shows the whey pathway in the plant. Raw whey is pasteurized and then sent to an ultrafiltration plant. The permeate fraction is then deashed by passage over an ion exchange demineralization unit. The demineralized whey is stored until further use. The hydrolysis plant operates at 360 liter/hr and achieves an 80% hydrolysis. The hydrolysis plant is entirely automatic and operates 16–20 hr/day.

Another pilot facility in France uses casein whey as a substrate; feed and product composition for pilot runs are shown in Table 2-1.

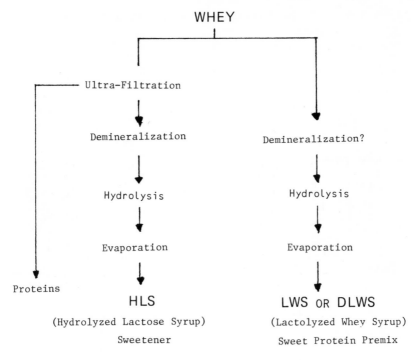

Figure 2-3.   Lactose hydrolysis route [7].

Table 2-1. Composition of the Casein Whey Permeate and Whey Permeate [7]

	Casein Whey Permeate[a] (demineralized)		Whey Permeate (partially demineralized)	
	Feed	Product	Feed	Product
Total Solids (%)	2.8	2.64	4.4	4.3
N × 6.38%	0.7	0.68	0.15	0.14
Ashes (%)	0.09	0.09	–	0.12
Lactose[b] (%)	2.6	–	4.1	–
Glucose (g/l)	–	11.4	–	16.4
Galactose	–	–	–	–
Total Counts	100-1,000	10^3-10^5	100-1,000	10^3-10^4
Yeasts	<10	10^3-10^5	<10-100	10^3

[a]Including casein wash waters.
[b]Lactose monohydrates.

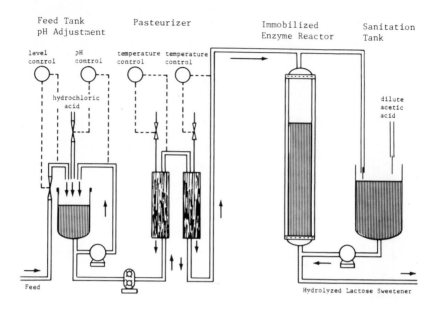

Figure 2-4.   General principle of operation [7].

Stage of Development

A semiindustrial plant exists at MMB Technical Division, Crudgington, England. There is a pilot-plant facility–Union Laitière Normande, Conde Sur Vire, France.

Implications for Energy Consumption

Relatively low energy requirements are needed for glucose manufacturing.

WASTE BIOMASS AS A SOURCE OF SINGLE-CELL PROTEIN (WATERLOO PROCESS)

Description

Production of SCP by fermentation has been widely commercialized all over the world. A major disadvantage of these various processes is their reliance on petroleum-based chemical feedstocks, which are now prohibitively expensive. The fermentation processing approach is also not a feasible option

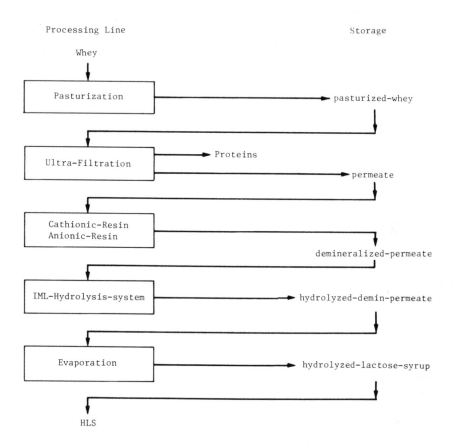

Figure 2-5.   MMB semiwork operations [7].

for many developing countries due to the high technology involved. A process described here, known as the "Waterloo Process," offers a potential solution to the above problems. The process utilizes raw materials such as agricultural, animal and forestry waste residues. Since this process uses waste by-products, it concurrently alleviates environmental pollution frequently generated by these wastes. This process employs low-technology operations with relatively small expenditures compared to the expenditures required for other fermentation processes.

The Waterloo process is based on the mass microbial cultivation of a new cellulolytic fungus, *Chaetomium cellulolyticum,* developed in the laboratory [8]. Figure 2-6 gives the generalized outline of the Waterloo bioconversion process for SCP production. The three-stage process involves: (1) thermal

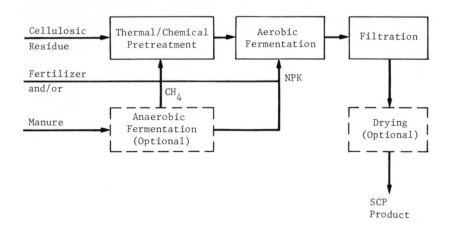

Figure 2-6. Generalized outline of the Waterloo bioconversion process for SCP production from agriculture or forestry waste [9].

and/or chemical pretreatment of a cellulosic material; (2) aerobic fermentation of the pretreated material with nutrient supplements; and (3) separation of the product from the fermented broth. The cellulosic material, which provides the principal carbon source, is pretreated with hot water or alkali, depending on the feedstock type. This reduces the inherent recalcitrance of the solids to fermentation by swelling and/or partial delignification. The basic process is carried out as a solid-substrate fermentation in slurry systems. The SCP product can be recovered by relatively simple, inexpensive filtration methods because of the mycelial nature of the fungus. The effluents of the process are essentially carbon dioxide and biochemical oxygen demand (BOD)-free water.

The basic Waterloo process given here has been improved to enhance productivity by using a polymer additive in the fermentation medium, a tubular fermenter design, semisolid systems and a mixed culture.

Table 2-2 presents an amino acid profile that is comparable with FAO (Food and Agriculture Organization of World Health Organization) reference standards, yeast and soybean meal. The SCP product of the Waterloo process is more attractive than the yeast SCP for direct human use because of its low nucleic acids content (5% vs 8% for yeast) and higher content of sulfur-containing amino acids. The low process costs are due to low-cost raw materials, process conditions (e.g., low temperatures and pressures) and efficient mass and energy exchanges between processing streams. The Waterloo SCP process has been tested successfully in a 200-liter pilot fermenter; further work is being undertaken to make technical improvements using a

Table 2-2. Essential Amino Acids (as % DM protein) in the Waterloo SCP
and Other Protein Products [9]

Amino Acid	FAO Reference	Soybean Meal	Waterloo SCP	Fodder Yeast
Isoleucine	4.2	4.2	4.7	5.3
Leucine	4.8	7.7	7.5	7.0
Lysine	4.2	6.4	6.8	6.7
Methionine + Cystine	4.2	2.2	2.6	1.9
Phenylalanine	2.8	4.7	3.8	4.3
Threonine	2.6	3.6	6.1	5.5
Tryptophan	1.4	1.7	N.A.	1.2
Tyrosine	2.8	2.7	3.3	3.3
Valine	4.2	4.4	5.8	6.3

continuous-flow, 1000-liter demonstration unit. This process can be made energy efficient if animal manure is used as the starting material and if methane gas, which is produced concomitantly, can be recycled.

Stage of Development

The technology is ready to be commercialized.

Implications for Energy Consumption

There are low energy requirements compared to other SCP production processes [9].

CHEESE PRODUCTION THROUGH LACTIC CULTURE SYSTEM (UTAH STATE UNIVERSITY)

Description

Conventional means of cheese production encounter a variety of problems such as bacteriophage infection, difficulty in bacterial cells adjusting to changes in pH, the high cost of casein and selection of a nitrogen source. The lactic culture system described here provides many economic advantages over conventional culture systems, according to its developers. The Utah State University (USU) Lactic Culture System combines conventional bulk culture tanks, pH and temperature controllers, and a diluted fresh liquid whey nutrient source supplemented with stimulants and phosphates/nitrates.

The bacteria produced in this system amount to two to five times the number of active cells per volume than with conventional cultures. In addition, the bacteria can be stored at 4°–13°C for days, thus eliminating the need for daily lactic culture propagation.

This commercial-scale USU Lactic Culture System requires sealed culture tanks providing slow-speed agitation throughout the culture fermentation cycle. In the United States, these tanks are not currently available. The tanks are also equipped with pH and temperature controllers. The streptococcal cultures are used to inoculate fresh liquid whey medium. The whey medium is supplemented with phosphates and some nitrogenous materials to make it inhibitory to phage development while providing maximum stimulation for the cultures. The lactose content of the whey medium is adjusted to 3.75% by diluting with water, thus ensuring a maximum growth of cells without any buildup of toxic end products.

Ammonia gas, which is less toxic to the cells than other neutralizers, is used to control the temperature of the system and the pH of the medium. When ammonia gas is injected, heat is generated, causing the culture temperature to rise to 27°C. The automatic temperature control mechanism starts operating whenever the temperature increases above 27°C.

First the bulk culture content in the tank is heated to 90°C for 45 minutes, providing a phage-free bulk culture. Then it is cooled to 27°C, followed by inoculation with *Streptococci* culture. The phosphate in the system provides buffering, and pH is maintained between 6.0 and 6.2. The pH control reduces bacterial generation times and increases the rate of cell production. When the pH is controlled between 6.0 and 6.2 less calcium is released. The more acidic the culture medium, the more calcium is dissolved. Since the pH of the medium is near neutral, a reduced amount of calcium is released, thus reducing the chances for the phage to attach to the bacterial cell wall.

The USU Lactic Culture System has lower equipment and operating costs than those associated with conventional phage inhibitory media systems. The USU system has been used successfully in the manufacture of American-style cheeses for over nine years, cottage cheese production for more than a year, and Swiss cheese production for six months [10]. The system has not been used for commercial production of Italian cheese, sour cream, buttermilk, yogurt or cream cheese.

Stage of Development

A commercial scale facility is in operation.

Implications for Energy Consumption

Potential for energy use reduction exists; this needs to be confirmed with cheese manufacturers using the system.

USE OF FUNGAL LACTASE IN DAIRY PROCESSING

Description

One of the best known lactase sources is the yeast *Kluyveromyces fragilis*. The limitations of this conventional source microorganism and certain other bacteria and fungi are as follows: (1) low activity; (2) loss of activity at elevated temperatures around 50°–60°C; and (3) slow cell growth, hence lower yield of lactase. In general, the uses of lactase in dairy processing are twofold: (1) increased sweetness in fluid milk without addition of sugars or artificial sweeteners; and (2) production of smoother ice cream and other concentrated milk products.

Thermophilic fungi capable of producing highly active thermostable lactase are currently in use. The organisms are grown on a culture medium containing lactose and glucose and, following successive cultures on diminished glucose and increased lactose concentration, the mycelium can be harvested. The separation and concentration of the lactase fraction is achieved via filtration, centrifugation, precipitation with polyacrylic acid, column chromatography using carboxymethylcellulose, gel exclusion chromatography using sephadex products, etc. This lactase source is stable and retains its activity at temperatures of 60°C or higher. The advantages of lactase derived from thermophilic fungi are as follows:

1. The time required for hydrolysis by use of high temperature is reduced, hence bacterial and fungal species responsible for inducing spoilage are inhibited.
2. Production of buttermilk, yogurt and sour cream can be achieved at low pH since lactase derived from thermophilic fungi exhibits maximum activity at a low pH (between 4-5).
3. Mixture of conventional lactase and lactase from thermophilic fungi can be used within a wide temperature range (low of 40°C and high of 60° or higher).
4. Lactase derived from thermophilic fungi aids in hydrolyzing lactase content of whole milk, whey or other dairy products to glucose and galactose, a sweetener product with no great caloric content.
5. Lactose removal by lactase producing fungi eliminates crystallization tendency in the manufacture of ice cream and other concentrated milk products (Figure 2-7).

Stage of Development

This process was patented in 1977.

Implications for Energy Consumption

1. Reduction in time required for milk processing.
2. Production of a variety of milk products concurrently based on different temperatures and pH requirements, e.g., buttermilk, yogurt at low pH.

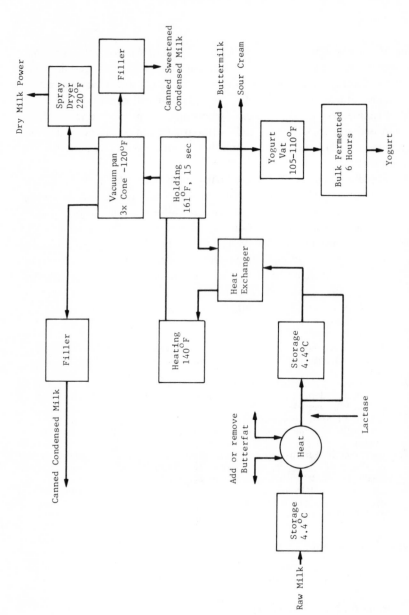

Figure 2-7.   Use of lactase from thermophilic fungi [11].

3. Whey as waste product can be treated at high temperature to produce glucose and galactose, thus overcoming whey disposal problem.

BACTERIAL LACTASE UTILIZATION IN DAIRY PROCESSING

Description

The enzyme lactase is used in the dairy industry to increase sweetness in fluid milk without the addition of sugars and sweeteners and to increase smoothness in ice cream and other concentrated milk products. Lactases have been derived commercially from lactose fermenting yeast, usually *Saccharomyces fugilis, Zygosaccharomyces lactis, Saccharomyces lactis* or *Candida pseudotropicalis.* Fungal preparations that have been found to be active as lactases under certain conditions are derived from *Aspergillus luchuensis, A. oryzae* and *A. niger.* The classic source of bacterial lactose is *Escherichia coli.* A major limitation to using lactases derived from these sources in processing dairy products has been low heat stability. They conventionally require long incubation (several days) at low temperature, or a relatively short (2–4 hours) incubation at 35–40°C.

The development of heat-stable lactase prepared from *Streptomyces coelicolor* has enabled production of lactase, which exhibits activity at a temperature of 70°C above the usual milk pasteurization temperature. This enzyme is produced by culturing *Streptomyces coelicolor* in the temperature range of 20°C to 50°C on a nutrient medium containing lactose under aerobic conditions. The growth medium also requires incorporation or proteinaceous material. The lactase prepared for *S. coelicolor* has its highest activity at an optimum temperature of 65°C, although 60°C to 70°C is an operable range with acceptable activity. At 65°C the enzyme activity is about three times greater than that at 40°C. This enzyme has been shown to have marked sensitivity to Cu^{++}, Fe^{++} and Ca^{++}. Among the metallic ions found in milk, copper appears to have the most harmful effect on lactase activity.

Stage of Development

This is available commercially as a subculture of *S. coelicolor.*

Implication for Energy Consumption

Pasteurization and lactase-associated enzyme action can be conducted simultaneously; thus a reduction in heat requirement and reaction time is a possibility.

USE OF IMMOBILIZED ENZYMES IN CHEESE PROCESSING

Description

Cheddar cheese manufacturing involves various steps, as shown in Figure 2-8. There are known to be two major drawbacks with this process:

1.   It is a batch rather than a continuous process.
2.   Proteolytic enzyme, which aids in coagulation of casein, is lost in the whey waste.

The coagulation stage (i.e., the cheese vat) is the main obstacle in converting from a batch to a continuous process. Here, an alternative is described that utilizes immobilized enzymes to coagulate the milk. The use of immobilized enzyme serves two purposes. First, it allows more efficient continuous process development. Second, it allows the recovery and reuse of the proteolytic enzyme.

The role of proteolytic enzyme in cheese manufacturing is twofold. It

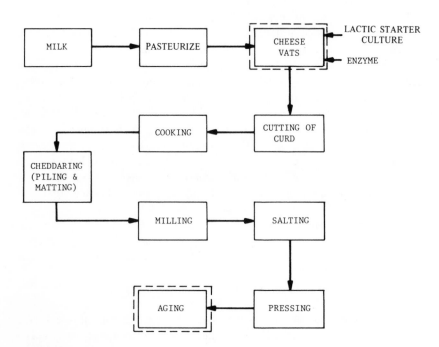

Figure 2-8.   Traditional process for cheddar cheese manufacture [12]. (Double-lined blocks indicate stages of enzyme action.)

causes coagulation by destabilizing the casein in milk and gives the characteristic cheese flavor resulting from low hydrolysis of protein and lipids. The primary phase of coagulation is characterized by the normal enzymatic reaction, while the secondary phase involves physical interactions. In the primary phase the system is sensitive to variation in pH and temperature, and here the application of immobilized enzymes in milk coagulation is most advantageous. Hence, by lowering the temperature of the reactor and/or increasing the pH of the milk, researchers were able to prevent clotting of the milk during passage through the immobilized enzyme reactor and also retain sufficient activity for the primary phase to occur. The third phase of gelation is accomplished by warming the milk and/or lowering milk pH.

In the laboratory experiment, proteolytic enzymes are immobilized on the most suitable catalyst support—glass. In the fluidized bed reactor, the immobilized enzyme is suspended with an upward flow of milk. This minimizes plugging and high-pressure drops in the reactor since the bed is free to expand with changes in fluid velocity (Figure 2-9).

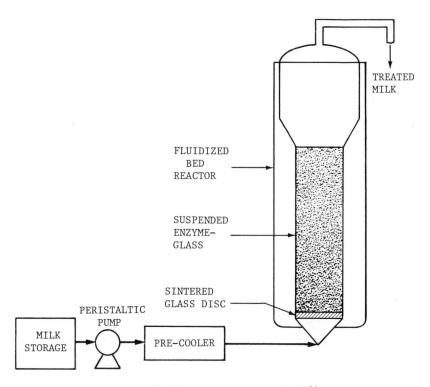

Figure 2-9. Diagram of enzyme reactor [13].

Using immobilized enzymes in a fluidized bed reactor has a number of advantages over conventional packed beds. Immobilized enzymes provide a more efficient continuous process for cheese manufacturing and also allow easier recovery and reuse of the enzymes, thus alleviating the critical shortage of commercial enzyme. Figure 2-9 illustrates an enzyme reactor. Pepsin or rennet is immobilized on alkylamine porous glass using glutaraldehyde as a coupling agent. Adaptation to continuous enzymatic milk coagulation has a potentially beneficial economic significance. At the normal pH of milk, the immobilized enzymes used for milk clotting are quite stable and active while the subsequent gelation occurs slowly. Thus, operating the reactor at the normal pH of skim milk, as well as lowering the temperature slightly, can retard coagulation significantly and increase the life of the catalyst.

Stage of Development

This is being used in commercial applications.

Implications for Energy Consumption

Potential energy reduction due to reduced operating temperatures.

WINE PRODUCTION FROM CHEESE WHEY WASTE

Description

Utilization of whey for wine production requires minimal energy resources. This is the advantage of the fermentation process described here in which wine is produced by the conversion of cheese whey. This technique has potential advantages for the U.S. cheese industry, which has been seeking development of a whey by-product which eliminates the use of expensive processes for water removal. The semiproduction scale process given here requires low capital outlay since no elaborate or expensive equipment is required.

The general scheme of wine production from whey is shown in Figure 2-10. The energy requirement for this process has been suggested to be 113 Btu/lb, approximately 950 Btu/gal (3.785 liters). Based on laboratory experiments with different fermentation yeast strains and temperatures, fermenting whey wine at room temperature appears to be the most energy-efficient approach. The results of these tests are given in Tables 2-3 and 2-4. Among the wine yeast strains tested, Montrachet strains were found to be the most desirable since the fermentation can be carried out at room temperature. Cheese whey itself also has been shown to provide sufficient nutrients for yeast growth during fermentation.

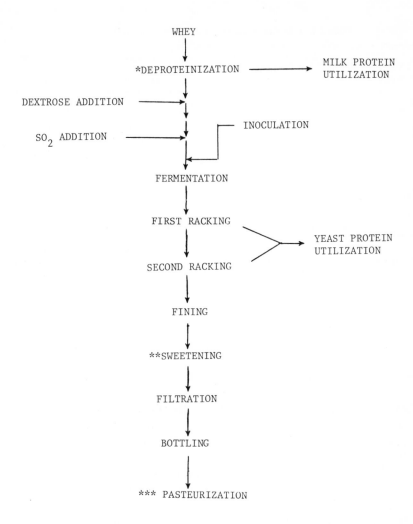

```
                          WHEY
                           │
                           ▼
              *DEPROTEINIZATION  ──────────▶  MILK PROTEIN
                           │                   UTILIZATION
                           ▼
DEXTROSE ADDITION  ───────▶│
                           │          ┌── INOCULATION
    SO₂ ADDITION  ────────▶│          │
                           │◀─────────┘
                           ▼
                    FERMENTATION
                           │
                           ▼
                   FIRST RACKING ──────┐
                           │            ▶  YEAST PROTEIN
                           ▼            ▶   UTILIZATION
                  SECOND RACKING ──────┘
                           │
                           ▼
                       FINING
                           │
                           ▼
                  **SWEETENING
                           │
                           ▼
                    FILTRATION
                           │
                           ▼
                      BOTTLING
                           │
                           ▼
              *** PASTEURIZATION
```

*Omitted if a cloudy wine is produced.
**Omitted when a dry wine is produced.
***Omitted if an aseptic method is used.

Figure 2-10.   Production of sweetened clean whey wine [15].

In the semiproduction scale demonstration, Jack cheese whey obtained from dairy producers is used at about pH 6.0. Following the removal of residual milk fat, clarified whey is supplemented with 22% added dextrose and 100 ppm sulfur dioxide. This whey is then placed in a 1500-liter stainless steel vertical dairy processing tank equipped with a slow sweep agitator. The

Table 2-3. Effect of Yeast Strain on Fermentation Rate [14]

Yeast Strain	Time for Completion of Fermentation (days)	Fermentation Rate per Day	Alcohol Production (%)
Montrachet	7	0.14	10.35
Champagne	8	0.13	10.25
Sherry	8	0.13	10.00
Port	12	0.08	10.20
Burgundy	14	0.07	10.10

Table 2-4. Effect of Temperature on Fermentation Rate [14]

Temperature (°F)	Time for Completion of Fermentation (days)	Fermentation Rate per day	Alcohol Production (%)
55	17	0.06	10.65
72	7	0.14	10.51
90	4	0.25	10.20

tank is cooled to 21°C using water. Active dry Montrachet wine yeast is then added at a rate of 12 g per 54 kg of supplemented whey; the agitator is started to disperse yeast; and the temperature is maintained at 20–22°C. The agitation is continued until active fermentation begins (approximately 24 hours), at which time mechanical agitation is stopped. The fermentation is essentially completed within 7–8 days, and at this time the alcohol content is between 9 and 10% by volume.

Deproteinization of cheese whey can be accomplished by heat denaturation of the heat-labile portion at 185°C, by chemical treatment using sodium pyrophosphate and calcium chloride followed by heating to 90°C, or by ultrafiltration. The by-product of this deproteinization is a whey permeate consisting of water containing lactose, milk minerals, nonprotein nitrogen, and miscellaneous vitamins and other minerals.

Following fermentation of the whey permeate, yeast biomass is removed by sedimentation followed by decantation, centrifugal separation and filtration, or "racking." A step of demineralization of fermented permeate is a necessary process step that yields a fermented whey beverage base (FWB) that is not too salty. Ion exchange or electrodialysis are the two methods that can be employed during the demineralization step.

Figures 2-11 and 2-12 illustrate process flowcharts projected to 4000-liter ferments for acid whey and sweet whey, respectively. The fermented whey beverage can be flavored by the addition of flavor, invert syrup and malic acid. The final product is known as flavored fermented whey beverage (FFWB). Currently, more research is in progress for flavor development.

The economics of fermented whey beverage processing have been studied and described in detail [16].

Stage of Development

This is a commercial project of Foremost Food Co., a Division of Foremost McKesson, Inc. Dublin, California.

Implications for Energy Consumption

Relatively low energy is required for this process due to low-temperature operation. It is useful utilization of cheese whey waste, which otherwise would require treatment.

CHEESE WHEY-DERIVED SWEETENERS

Description

Whey is a liquid by-product from cheese manufacturing and contains 0.8% soluble protein of high nutritive quality, about 5% lactose, various minerals and other constituents. The lactose in whey constitutes a utilization or disposal problem. The method described here facilitates lactose conversion to glucose and galactose by enzymatic hydrolysis. Lactase is an enzyme that is produced by the organism *Aspergillus niger* and is immobilized by adsorbing on a porous alumina carrier followed by a mild cross-linking with glutaraldehyde.

This lactose catalyst has been employed in a pilot-plant hydrolysis unit at the Lehigh Valley Dairy in Allentown, Pennsylvania. The unit consists of two 6-foot (1.8 meter) columns of 3-inch (7.6 cm) internal diameter, each containing about 4 kg of immobilized lactose catalyst and having a maximum daily throughput capacity of about 40 gallons (151.4 liters) of whey. These columns are equipped with conical inlet sections above quick-acting ball valves. This modification prevents clogging of columns by untreated whey, can provide good liquid distribution and minimize channeling during operation [17]. The flow diagram of hydrolysis reactors and associated operating

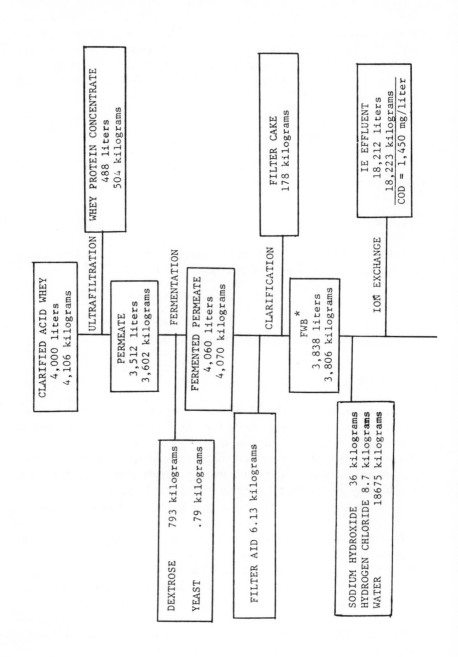

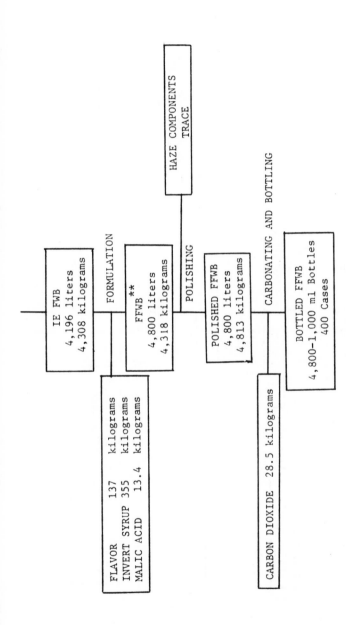

Figure 2-11. Process flow diagram for FFWB from acid whey [16].

*FWB Whey Beverage
**FFWB Flavored Fermented Whey Beverage

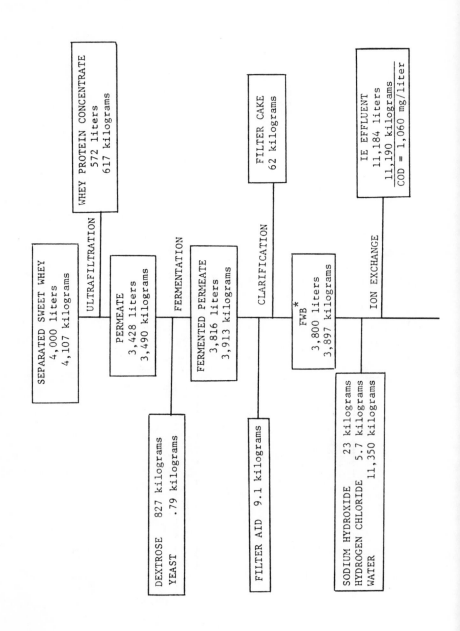

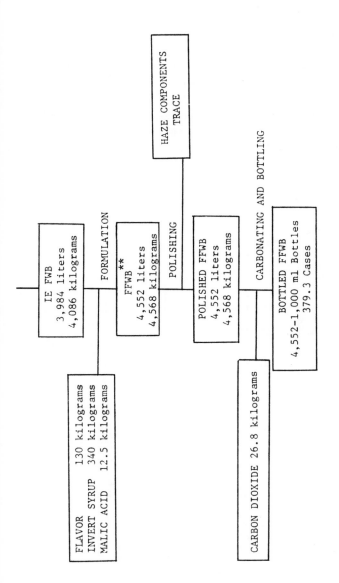

Figure 2-12.   Process flow diagram for FFWB from sweet whey [16].

*FWB Fermented Whey Beverage
**FFWB Flavored Fermented Whey Beverage

units for protein separation, demineralization and concentration is given in Figure 2-13.

The raw whey is first heated to the desired hydroloysis temperature (45°-60°C) in a stainless-steel shell and tube heat exchanger. The whey is then pumped through the fluidized-bed, immobilized enzyme hydrolyzers, which convert the lactose in whey to glucose and galactose. The hydrolyzed whey then flows through a particle trap to remove any immobilized catalyst that might elutriate (separate out). In a liquid fluidized bed very little of the catalyst is elutriated since this phase is characterized by gentle agitation of the catalyst particles and a sharp liquid catalyst interface at the top of the bed. The demineralization step in this pilot process serves as a pretreatment of the whey. Pretreatment of whey is also possible by subjecting whey to ultrafiltration alone or in combination with demineralization.

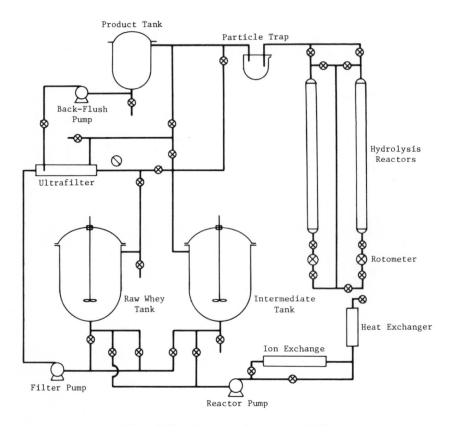

Figure 2-13.   Pilot plant flow diagram [18].

It has been demonstrated that the conversion of 5% lactose is more efficient with this pilot operation than with a bench-scale operation. Table 2-5 shows comparative data for the pilot-plant versus bench-scale operation for hydrolysis of lactose solutions. A 64% conversion of lactose to glucose can be accomplished in about 30 minutes in the pilot-plant reactor.

Stage of Development

There is a pilot plant at Lehigh Valley Dairy, Allentown, Pennsylvania.

Implications for Energy Consumption

This process is energy efficient, and estimated operating costs are relatively low for the processes of hydrolysis, concentration, demineralization, ultra-filtration, etc. [17].

PROCESSING OF CORNSTARCH TO PRODUCE SWEETENING AGENTS (PILOT)

Description

In the application discussed below, starch (e.g., corn) is processed to fructose by enzymatic hydrolysis using a bacterial *(B. licheniformis)* amylase,

Table 2-5. Comparison of Pilot-Plant and Bench-Scale Results
for Hydrolysis of Lactose Solutions [18]

	Bench Scale	Pilot Plant
Reactor Diameter (cm)	2.54	7.56
Unexpanded Bed Height (cm)	10.00	96.00
Expansion (%)	60%	40%
Catalyst Weight (g)	50%	3900
Catalyst Activity (LU/g)[a]	4400	6500
Residence Time (LU-min/ml)	5.0%	507×10^2
Lactose Concentration (wt%)	0.06	5.0%
Flowrate (liter/min)	65%	0.50
Conversion (%)		84%

[a]Catalyst activity assayed at 37°C, based on catalyst activity at 37°C.

which is stable at high temperature. This process was introduced in 1973 by NOVO Industries.

The process can be represented by three stages: liquefaction, saccharification and isomerization. Table 2-6 gives the physical parameters of starch processing. Enzyme liquefaction produces higher yields of glucose than with conventional acid liquefaction. The best results are obtained by employing a two-stage enzyme process consisting of thinning and cooking, followed by saccharification, as shown in Figure 2-14. Heat-stable "TERMAMYL®" can be used with the existing equipment designed for acid liquefaction and can be adapted at minimum investment [19]. A 30–40% dry solid starch slurry is prepared at pH 6.5–7.0. Liquid enzyme, *B. subtilis* amylase, is fed directly into the starch slurry stream immediately before the converter. The temperature is raised to 105-110°C by direct, live stream injection. The starch slurry is retained in the converter coil for 5-7 minutes at high temperature. The liquefied starch is then flashed to 95°C and held for 60–90 minutes to complete liquefaction. The advantages of using an enzyme that is stable at high temperature are (1) reduced operating costs, as less energy is expended in pressure cooking, and (2) low viscosity, reducing the power required for stirrers and gearboxes. In addition, this process does not require calcium ion for enzyme activity enhancement, as in the case of conventional α-amylase.

The saccharification step is similar to that in conventional starch conversion processes. There, the amylose fraction of liquefied starch is broken down into glucose by the action of amyloglucosidase.

The step of isomerization is illustrated in Figure 2-15. Here, syrup from the saccharification step is first purified by centrifugation/filtration and then carbon filtered. The syrup is then evaporated to a percent solids level of 40–45% dry substance content and ion exchanged to remove all cations.

Table 2-6. Physical Parameters of Starch Processing [19]

Parameter	Liquefaction	Saccharification	Isomerization
Enzyme	Termamyl	Amyloglucosidase	Sweetzyme
Temperature (°C)	107	60	65
pH	6.5	4.5	8.5
Solids (%)	40	35	40-45
DE	12	98	–
Activators	0	0	Mg
Reaction time (hr)	2	24-76	2

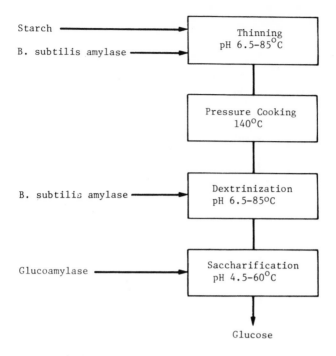

Figure 2-14.   Enzyme-enzyme liquefaction.

Magnesium is added as an activator, and the pH is adjusted to alkaline conditions. The feed syrup is passed to the isomerization columns, warmed to 65°C and circulated in the columns through a protective filter. After isomerization, the syrup is pH-adjusted back to acidity and either passed forward to confection or evaporated for storage. This isomerization step uses insolubilized enzyme; the advantages of such an enzyme are as follows [19]:

• multiple reuse
• lack of residual enzyme in the finished product
• improved stability
• modified catalytic properties
• easier processing and control
• multiple operation modes
• low operation cost

The pH optimum of the isomerization step is around 8.5 and the temperature optimum is around 85°C, although reactors usually are run at 65°C. The reason for this is that 65°C is a compromise temperature for maximum enzyme stability and minimum endogenous by-product formation from the produced fructose. Continuous isomerization can be achieved by the use of

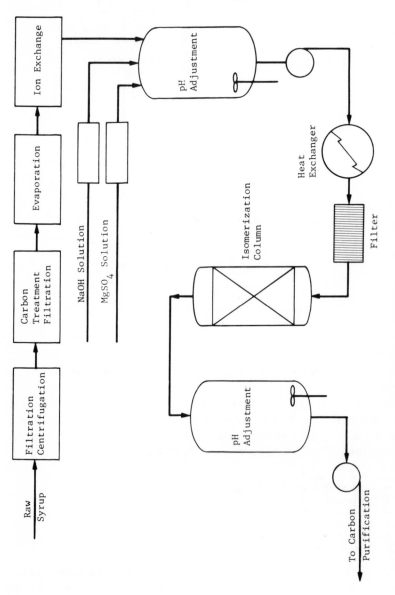

Figure 2-15.   Continuous isomerization [19].

multiple columns. The glucose produced via enzyme utilization is further converted to fructose by isomerization using glucose isomerase enzyme. Table 2-6 summarizes the physical parameters of starch processing.

Stage of Development

There is a pilot plant, at Novo Laboratories. The enzyme TERMAMYL is commercially available.

Implications for Energy Consumption

Potential reduced energy consumption is achieved by the reduction in time required for pressure cooking and reduction in use of stirrers and gearboxes.

SWEETENER (FRUCTOSE SYRUP) PRODUCTION FROM CORNSTARCH (COMMERCIAL)

Description

The basis for the successful development of fructose syrup products is the discovery of the enzyme glucose isomerase, also known as D-xylose isomerase. This enzyme is produced in most microorganisms capable of growing on xylose sources. Table 2-7 lists the organisms presently being used and the companies producing enzymes. Glucose isomerase is generally produced commercially by submerged aerated fermentation in several stages. The three development stages of such a production technique are described in U.S. Patent 3,666, 628 [20]: (a) slant development; (b) culture development—two substages; and (c) final fermentation stage. Following recovery from the microorganisms, the enzyme is immobilized by one of a number of techniques (see Antrim et al. [21] for listing). Glucose isomerase is used in the manufacture of fructose syrup from cornstarch, which is the most widely available and economical source of dextrose.

Cornstarch manufactured by a corn wet milling process is by far the most widely used raw material for dextrose production in the United States. Wet milled cornstarch is comprised of a polymer containing about 99% anhydro-dextrose on a dry basis. Besides corn, wheat could also be a logical choice in major wheat-producing areas of the United States, Australia and Canada.

The use of enzymes has largely replaced conventional acid-catalyzed liquefaction and saccharification of starch. A typical process for dextrose manufacture from cornstarch for use in making fructose corn syrup is given in Figure 2-16. Starch slurry at about 33% dry solids is liquefied with a bacterial L-amalyse at temperatures ranging from 80° to 110°C. The liquefaction

Table 2-7. Available Commercial Immobilized Glucose Isomerase
Preparations and Enzyme Sources [21]

Firm	Enzyme Source	Immobilization Procedure	Enzyme Form
Clinton Corn Processing Company	*Streptomyces ribigenosus*	Adsorption on anionic celluloses and composites	Fibrous and granular
Novo Industri	*Bacillus coagulans*	Lysed cells cross linked with glutaraldehyde	Granular
Gist Brocades	*Actinoplanes missouriensis*	Whole cells entrapped in glutaraldehyde cross-linked gelatin	Granular
ICI Americas, Inc.	*Arthrobacter*	Flocculated whole cells	Granular
Miles Labs, Inc.	*Streptomyces olivaceus*	Glutaraldehyde cross-linked whole cells	Amorphous
CPC Int. Inc.	*Streptomyces olivochromogenus*	Adsorption on alumina (porous) or other ceramic carriers	Granular
Nagase	*Streptomyces phaeochromogenes*		Granular
Miles-Kali Chemie	*Streptomyces*	Fixed cells	Amorphous
Sammatsu	*Streptomyces*	Adsorption on anion exchange resin	Granular

process is continuous, with residence times ranging from 2 to 4 hours. The α-amylase used has a calcium requirement (200–500 ppm, dry basis), and the pH is adjusted to 6-7 with calcium hydroxide.

After flash-cooling to about 60°C, the liquefied starch is saccharified by treatment with a fungal glucoamylase. Conditions for saccharification are 55°-60°C, pH 4-4.5 and a holding time of 24-90 hours, depending on the amount of glucoamylase used and production scheduling. This is followed by steps of filtration, refining, isomerization and concentration to yield 42% fructose syrup containing no more than 6% nonmonosaccharides. The reactors that are used for the isomerization step are of six different design types: batch enzyme, packed-bed, continuous-flow stirred-tank, continuous-flow stirred tank/ultrafiltration membrane, recycle reactors and tubular reactors with enzymatically active walls. Pretreatment of dried, immobilized glucose isomerase before introduction into a reactor is generally recommended as a necessary procedure for minimizing pressure drop, as well as for attaining high levels of enzyme activity. Pretreatment accomplishes two things:

1. It hydrates and swells the enzyme particles, thus preventing compaction.
2. It equilibrates the pH at or near operating pH.

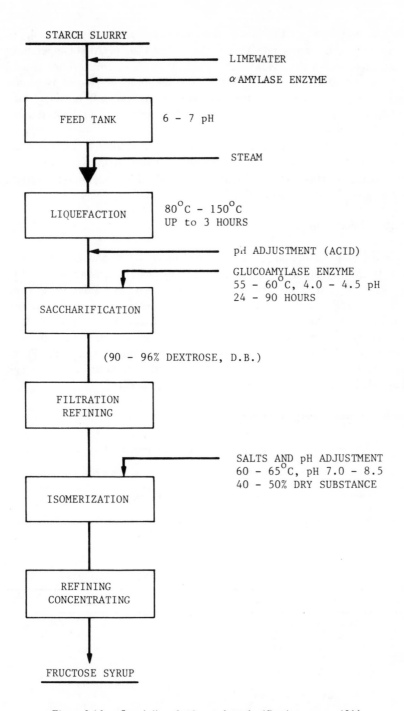

Figure 2-16.   Starch liquefaction and saccharification process [21].

Glucose isomerases in the immobilized form are reasonably stable up to about 70°C, and are generally unstable at commercial operation temperatures in substrate with a pH less than about 6.5. They also seem to be inactivated to a certain extent by air oxidation. The immobilized glucose isomerase enzyme activators include magnesium ion, citrate ion, sulfurous acid salts and ferrous ions. The enzyme is known to be inactivated by microbial infection of reactor streams.

The high-fructose syrup recovered following isomerization is carbon-treated and deionized with a strong acid cation exchange resin. The process of refining following isomerization removes salts added for enzyme stabilization, and adjusts pH and traces of color.

Stage of Development

Commercial Production. Table 2-8 lists companies that manufacture 42% fructose corn syrup in the United States. In Europe, particularly in the EEC, development of fructose syrup production has been limited due to the competing interests of the sugar industry and a subsidy on exported beet sugar [21]. Another constraining factor is that European corn is difficult to process by wet milling. Since 1976, fructose syrup plants have been in operation in Belgium, Germany, the Netherlands, Britain and Spain. Plants are under construction in France (Société des Produits du Maise and Roquette Frères), Ireland, Italy and the Netherlands [21].

Implications for Energy Consumption

1. Process of corn wet milling, which requires drying by evaporation later in the process, is energy intensive.
2. Use of more efficient evaporators (e.g., mechanical high-efficiency evaporators) might mitigate this energy consumption.
3. Energy requirements for refining and cleaning are minimized by biotechnology.
4. Small, flow-through reactors use less energy.

Table 2-8. Producers of 42% Fructose Corn Syrup (U.S.) [21]

Company	Brand Name
American Maize Products Company	TruSweet®
Amstar Corporation	Amerose
Archer Daniels Midland Corn Sweeteners	Corn Sweet®
Cargill, Inc.	ISOCLEAR
Clinton Corn Processing Company	ISOMEROSE
CPC International, Inc.	INVERTOSE®
The Hubinger Company	HI-SWEET
A. E. Staley Manufacturing Co.	ISOSWEET

HYDROCARBON-DERIVED YEAST PROTEIN FOR
HUMAN CONSUMPTION

Description

Yeasts have been a component of human food since ancient times. Present-day processes for producing yeast (food or feed) protein originated in Germany during the two world wars. *Saccharomyces cerevisiae* and *Candida utilis* were grown on molasses and sulfite-waste liquor, respectively. Table 2-9 presents a summary of some of the more important large-scale yeast production processes.

Two types of substrates have been used in these processes: (a) hydrocarbons, such as purified n-alkanes, gas, oil and kerosene, and petrochemicals derived from hydrocarbons such as ethanol; and (b) agricultural, industrial and food-processing wastes containing carbohydrates. Figures 2-17 and 2-18 show a schematic diagram for producing yeasts, such as *Candida* sp., from purified alkanes and carbohydrates, respectively.

Ethanol also has been used for large-scale production of food or feed yeast. *Candida utilis* is produced from purified ethanol by a process that operates continuously under sterile conditions to yield cell densities in the range of 6 to 7 g/l (dry weight basis). This product meets U.S. Food and Drug Administration regulations for food yeast and is being sold as a functional food additive. Figure 2-19 illustrates the "Pure Culture Process" of Amoco Foods Co.

Stage of Development

A commercial-scale facility is operated at Amoco Foods Co., Pure Culture Products, Inc. (Affiliate of Amoco Food Co.), Chicago, Illinois.

Implications for Energy Consumption

The energy requirement for *Candida* sp. grown on ethanol is estimated to range from 185 to 190 MJ/kg of protein [24].

HYDROCARBON-GROWN YEAST AS SUPPLEMENTAL
PROTEIN IN ANIMAL FEED

Description

It has been indicated that hydrocarbon-grown yeasts are likely to be used to replace wholly, or in part, the more conventional proteinaceous components

Table 2-9. Processes for Microbial Protein Production from Yeast [22]

Process	Scale	Substrate	Organism	Status
British Petroleum Co. Ltd., Grange-mouth, Scotland	4,000 metric ton/yr	Purified n-alkanes	*Candida lipolytica*	Pilot plant operation discontinued
Lavera, France	16,000 metric ton/yr	Gas oil	*Candida* sp.	Pilot plant discontinued
Sardinia	100,000 metric ton/yr	Purified n-alkanes	*Candida* sp.	Plant constructed but not operated
Liquichimica Biosintesi s.p.a. & Kanegafuchi Chemical Industry Co. Ltd., Reggio, Calabria, Italy	100,000 metric ton/yr	Purified n-alkanes	*Candida novellus*	Plant constructed but not operated
All-Union Research Institute of Biosynthesis of Protein Substances, USSR	20,000 & 40,000 metric ton/yr	n-Alkanes	*Candida* sp.	Plant constructed
Symba Process Svenska Socker Fabriks, Areov, Sweden	500 m^3/day	Potato waste	*Saccharomycopsis fibuligera* and *Candida utilis*	Waste treatment operation
Amoco Foods Co., Standard Oil Co. of Indiana, Hutchinson, MN	5,000 ton/yr	Ethanol	*Candida utilis*	Plant in operation
Amber Laboratories Div., Milbrew, Inc., Juneau, WI	5,000 ton/yr	Cheese whey	*Kluyveromyces fragilis*	Plant in operation
Boise-Cascade Corp., Salem, OR	5,000 ton/yr	Sulfite waste liquor	*Candida utilis*	Plant in operation

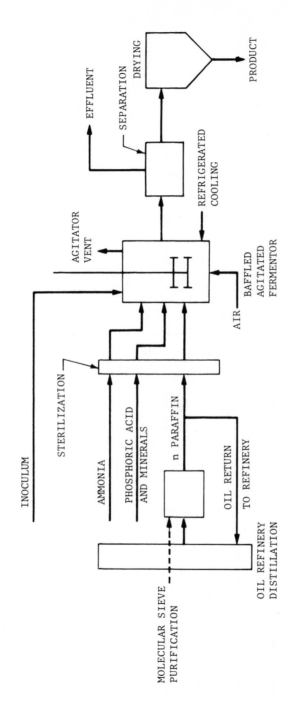

Figure 2-17.   Process diagram for producing yeast from purified n-alkanes[22].

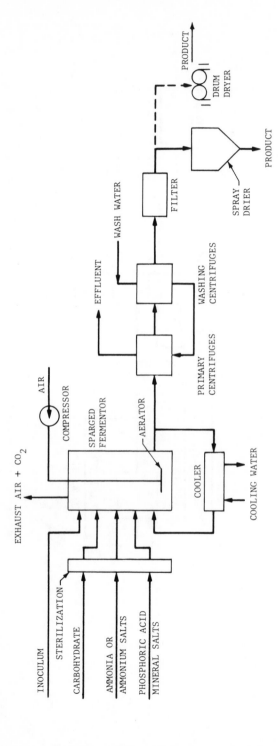

Figure 2-18. Process diagram for producing yeast from carbohydrates [22].

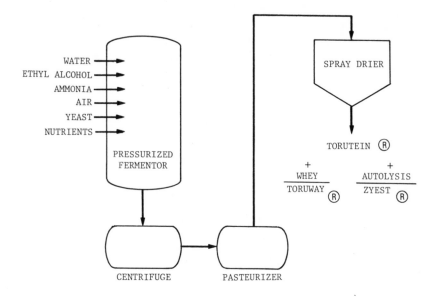

Figure 2-19.   Pure culture process [23].

of mixed animal feeds for pigs and poultry. The potential of hydrocarbon-grown yeast as a protein-rich component of mixed feed can be estimated from the quantities of the essential amino acids present. Table 2-10 is a comparative analysis of selected amino acids of yeast. Such a mixed feed is currently being tested for toxicity in experimental animals.

Large-scale production of yeast is possible using hydrocarbons such as liquid n-paraffins, which have low fuel value. Table 2-11 lists selected patents published in recent years, which are indicative of the worldwide interest in the commercial potential of biomass production from hydrocarbons. As of 1974, three small-scale industrial plants were in operation, two operated by the British Petroleum Co., Ltd., and a unit operating in the USSR. Other plants are under construction or planned elsewhere in the world.

Continuous fermentation is the most efficient and economical means of harvesting yeast from hydrocarbons [25]. Flow diagrams of the two British Petroleum processes are shown in Figures 2-20 and 2-21 and are typical of production routes published elsewhere. Figure 2-20 illustrates an aseptic process based on a purified (purity greater than 95%) n-paraffins feedstock, while Figure 2-21 shows the flow diagram of a gas oil process under non-aseptic conditions. The conditions required to develop economically viable processes include the following:

Table 2-10. Comparative Analyses of Selected Amino Acids of Yeast (g/16 g nitrogen) [25]

Amino Acid	Reference Proteins		Carbohydrate-Grown Yeast		Hydrocarbon-Grown Yeast		
	FAO Standard (323)	Whole Egg (174)	S. cereviseae (174)	Candida yeast 'Torula' (238)	Candida lipolytica (327)	Candida lipolytica (251)	Candida tropicalis (327)
Isoleucine	4.2	6.7	4.6	5.3	5.1	2.0	5.3
Leucine	4.8	8.9	7.0	7.0	7.4	2.8	7.8
Lysine	4.2	6.5	7.7	6.7	7.4	3.3	7.8
Phenylalanine	2.8	5.8	4.1	4.3	4.3	1.9	4.8
Tyrosine	2.8	4.2	–	–	3.6	2.0	4.0
Cystine	2.0	2.4	–	–	1.1	–	0.9
Methionine	2.2	3.2	1.7	1.2	1.8	0.4	1.6
Threonine	2.8	5.1	4.8	5.5	4.9	2.2	5.4
Tryptophane	1.4	1.6	1.0	1.2	1.4	–	1.3
Valine	4.2	7.3	5.3	6.3	5.9	2.2	5.8

Hydrocarbon-Grown Yeast

Amino Acid	Candida tropicalis (191)	Candida tropicalis (286)	Candida sp. (77)	Pichia guilliermondii (278)	Pichia sp. (69)	Trichosporon pullulans (339)	'Yeast' (359)
Isoleucine	5.8	2.5	4.8	4.3	9.5	6.0	5.5
Leucine	6.8	3.4	8.2	6.4	11.1	7.4	6.9
Lysine	10.4	3.5	10.7	7.6	13.9	6.2	7.5
Phenylalanine	2.3	2.2	4.3	3.8	9.4	3.9	4.4
Tyrosine	–	1.0	–	3.0	4.7	–	–
Cystine	–	0.1	–	0.7	1.5	–	1.8
Methionine	0.6	0.4	0.7	1.4	2.3	1.9	1.2
Threonine	5.5	1.8	3.2	5.8	6.3	4.3	5.1
Tryptophane	4.3	0.4	1.7	0.8	2.0	1.0	1.2
Valine	5.1	2.6	6.3	5.0	7.7	7.1	5.4

Table 2-11. Selected Patents Concerning Yeasts Grown on Hydrocarbons [25]

Company (or Author)	Reference[a]	Feedstock[b]	Organism
Asahi Chemical Ind. Co., Ltd.	24	n-Paraffins	*Candida lipolytica, C. tropicalis*
	28	n-Paraffins	*C. petrophilum, Brettanomyces*
	29	Wide range n-Paraffins	*petrophilum, Torulopsis petrophilum*
Atlantic Richfield Co.	31	n-Alkanes	*Candida intermedia*
British Petroleum Co., Ltd.	50	Gas oil	*C. tropicalis* (aqueous recycle)
	51	Gas oil	*C. lipolytica* (droplet size)
	52	n-Paraffins	*C. lipolytica* (hydrocarbon limiting)
	53	Gas oil	*C. tropicalis* (two-stage fermentation)
	54	n-Paraffins	*C. lipolytica* (overpressure)
Chepigo, S. V., et al.	65	n-Paraffins	*C. guilliermondii*
Czechoslovakian Academy of Science	71	n-Paraffins	*Candida* sp.
Deutsch Akademie Wissenschaft, Berlin	78	Hydrocarbons	*C. lipolytica*
Imada, O. and Hirotani, S.	150	n-Paraffins	*C. zeylanoides*
Institut Français Petroleum	152	n-Paraffins	*C. lipolytica*
Kanegafuchi Chemical Ind. Co., Ltd.	164	n-Paraffins	*Pichia* sp.
Kyowa Hakko Kogyo Co., Ltd.	194	n-Paraffins	*Torulopsis famata, C. zeylanoides*
	196	Hydrocarbons	*C. lipolytica*
	198	Liquid HC gas	*C. ridida, C. japonica*
	205	Hydrocarbon	*C. lipolytica, C. tropicalis, C. zeylanoides, Torulopsis famata, T. intermedia*
Rodionova, G. S., et al.	313	n-Paraffins	*C. guilliermondii*
Takeda, L., et al.	365	Hydrocarbons	*C. lipolytica*
Texaco Petroleum Co., Ltd.	376	Gas oil	*C. guilliermondii, C. rugosa*
Toyo Koatsu Ind. Inc.	380	n-Paraffins	*Trichosporon japonicum*
Uprav. Prom. Blek. Ves. Fer.	393	n-Paraffins	*Candida guilliermondii*

[a]Refer to the article by Shennan and Levi [25].
[b]Original nomenclature.

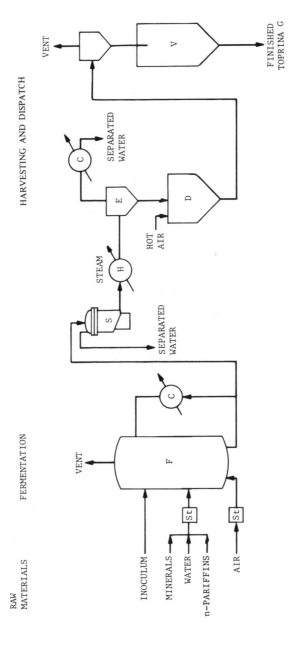

Figure 2-20.   Flow diagram for the BP n-paraffins protein process. (Symbols same as in Figure 2-21.)

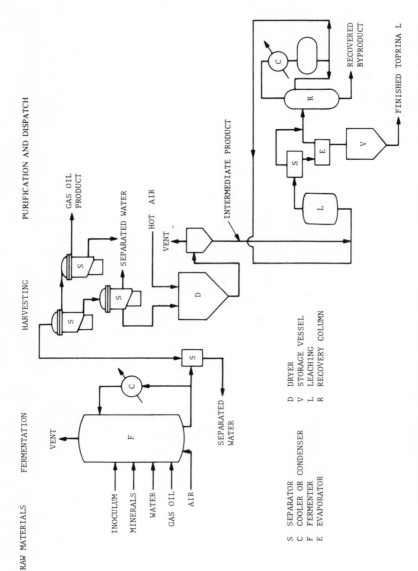

Figure 2-21.   Flow diagram for the BP gas oil protein process developed at Lavera [25].

- sufficient energy to disperse the hydrocarbon finely into the aqueous phase so that hydrocarbon mass transfer to the cell is not the rate-limiting step,
- large quantities of atmospheric O_2 transferred to the cells and similar quantities of CO_2 evacuated,
- low-grade heat eliminated at about 30°C, and
- food industry standards of hygiene to safeguard final product quality.

The energy requirements of these processes are relatively high. The precise power requirement is closely linked both to the mechanical design of the fermenter itself and to the microorganisms used, insofar as they can affect the methodology (flow) of the broth. The microorganisms concentration in the fermenter broth is critical in terms of the energy used during fermentation. The cooling is equally critical since hydrocarbon oxidation is a highly exothermic process. A cooling water requirement of 1200–1500 ton/ton dry weight of yeast is estimated. This demonstrates the serious limitation that a 30°C fermentation temperature represents. Such a process temperature would not be possible without refrigeration. A microorganism with an optimum growth temperature of 40°–45°C is desirable. Yeast species used in current commercial processes and pilot experiments have optimum growth temperatures generally in the range of 26° to 34°C. Work is underway to study the use of thermophilic yeast strains that can withstand higher growth temperatures of 40°-45°C [25].

Stage of Development

Commercial. There is ongoing pilot-stage research and development.

Implications for Energy Consumption

It requires substantial energy for fermentation as well as cooling; use of thermophilic yeast strains would minimize or reduce current energy requirements.

SINGLE-CELL PROTEIN THROUGH MASS ALGAL CULTURE

Description

The intensive mass cultivation of algae offers almost limitless protein production possibilities, as yields of 50% protein per unit dry weight and more have been achieved. Of all the microbial protein production processes, algae require the least technological sophistication [26]. The requirements for good algal growth include a supply of CO_2, nutrients, illumination, maintenance of optimal temperature, and adequate agitation to prevent settling of cells and to ensure an even exposure to CO_2, nutrients and illumination.

Algal mass culture production on waste substrates requires $1/50$ of the land area, $1/10$ of the water, $2/3$ of the energy, $1/5$ of the capital and $1/50$ of the human resources to produce equivalent amounts of food compared with conventional agriculture [26]. The process developed in South Africa utilizes the green alga *Chlorella vacuolata* as a test organism (inoculant). The chlorella culture (5 liters) is used to inoculate 200-liter outdoor batch cultures. The laboratory grown cultures are first allowed a three-day adaptation period under outdoor conditions before starting the mass cultures.

Oval fiberglass basins, measuring 1 × 2 meters with a surface area of 1.78 m^2 and 0.2 meters deep are used to culture the algae. The basins have a central partition or island, around which the culture is stirred at about 25 meters/min (82 ft/min) by an aluminum-stainless steel paddle wheel, turning at 24 rpm. Each basin is equipped with a CO_2 dosing system and an outlet. The algal growth medium consists of nutrient medium supplement with fertilizers in particular areas and superphosphates. Algal growth occurs in the system at outdoor temperatures above 10°C.

Based on the laboratory test results, the mass cultivation of algae seems feasible. Using the 1978 South African prices of fertilizers (R226.80 per metric ton urea and R66.25 per metric ton superphosphate) 1 kg of algae can be produced for approximately 10 ¢. In other words, in 1980 U.S. dollars, prices of fertilizers are estimated to be $311 per metric ton urea and $88 per metric ton superphosphate (at conversion rate of R1 = $1.3 U.S.). It is estimated that a production cost of R200 per metric ton ($266 U.S.) of algal protein can be achieved for average protein content of 50% in the product.

Stage of Development

Laboratory-scale research and development is underway.

Implications for Energy Consumption

There is an estimated $2/3$ reduction of energy use while producing algal proteins when compared with conventional crop production.

PRODUCTION OF STARTER CULTURES (COMMERCIAL)

Description

In many countries, conventional starter bacterial cultures for cheese-making are now being superseded by concentrated starters. Concentrated starters are prepared on a large scale at a centralized production facility, where strain characteristics, phage resistance, viability and acid production are rigorously controlled to produce uniform, high-quality starters. Use of the concentrated starters avoids the need for subculturing in the laboratory and propagation of bulk starter. A technique called direct vat inoculation has been suggested to make use of these concentrated starters rather than conventional bulk starters [27].

An important factor in retarding milk fermentation is lysis (cellular breakdown) of the lactic acid culture bacteria by bacteriophage (viruses that attack bacterial cells). The relationship between the lactic *Streptococci* and their phages influences most milk fermentations. In a given cheese processing plant, up to 10^{18} bacterial cells and 10^{16} phage particles may be produced each day. It is difficult to differentiate between cells and phages.

Long before phages were known to exist, it was common practice for cheese-makers to hold back each day some sour milk to "start" the next day's cheese-making. About 80 years ago commercial firms began to supply these starter cultures. Over the years these cultures have changed as a result of the interaction between bacteria in the milk and their phages.

Various mesophilic lactic *Streptococci* cultures are used in the manufacture of such fermented milk products as cottage, cheddar, camembert, blue and swiss cheese as well as yogurt. Commercial milk fermentations are usually batch culture operations and are essentially anaerobic. At the optimum growth temperature of 30°C, lactic *Streptococci* have mean generation times in milk of 60–70 minutes and grow to a maximum cell density of about 0.5 mg dry weight bacteria per milliliter of milk. The growth of lactic *Streptococci* in milk stops as a result of the lactic acid developed (pH 4.5). The longer the cells in these stationary-phase cultures are exposed to high acid concentrations, the more prolonged is the lag on subculture. During the manufacture of most fermented milk products, bacterial growth takes place with cells embedded in a gel. Hence, nutrient availability and end product dispersion are restricted within microcolonies of bacteria.

The manufacturing procedures for starter cultures in cheese-making are varied, but traditionally three stages of culture propagation are involved (Figure 2-22) [28]. Recent innovations in culture technology have centered on the replacement of liquid stock cultures, traditionally prepared at the commercial plants by frozen concentrated cultures provided either by dairy research establishments or starter supply companies. The cultures are grown in a suitable medium, preferably with pH control, frozen and then stored at temperatures no higher than -35°C. The survival of frozen cultures is influenced by the rate of freezing and the temperature of storage. The cell number in these frozen cultures concentrates usually ranges from 10^{10} to 10^{11} cells per gram. Centrifugation is the concentration process used currently, but a pilot-scale diffusion culture technique is reported to give a

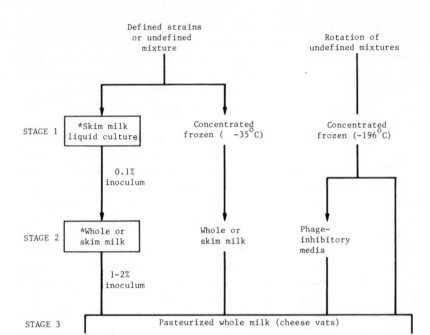

Note: Blocked Stages indicate traditional procedure

Figure 2-22.   Culture systems used in the manufacture of cheddar and gouda cheese – stages of propagation [28].

twentyfold concentrating effect during growth [27]. Another method of much interest uses lyophilized culture concentrates to avoid the cost involved in transporting frozen cultures.

The advantage of addition of concentrated cultures directly to the vats is that it decreases the risk of phage infection, since two of the three traditional culture propagation steps are eliminated. The second advantage of the direct vat inoculation is convenience, for which the cheese-maker and, thus, eventually the consumer, must pay. Concentrated starters are used primarily for bulk starter in cheese-making.

NOTE: Commercial cultures (concentrated and/or frozen) are available from various culture producers: (1) Fargo Series, Microlife Techniques, Sarasota, Florida; (2) Chr. Hansen's Laboratory Inc., Milwaukee, Wisconsin.

Stage of Development

Commercial production is underway [27-29].

Implications for Energy Consumption

None apparent at present.

USE OF MIXED BACTERIAL CULTURES IN YOGURT PRODUCTION

Description

Continuous yogurt production (cultivation of lactic acid-producing micro-organisms) can be successfully achieved by the use of a fermenter that has a primary application in milk-based media. The laboratory-scale, pH-Stat continuous fermenter is shown in Figure 2-23. A pH-Stat fermenter is a continuous cultivator in which the feedrate is controlled to maintain a constant pH, i.e., end product acid concentration.

Yogurt is made by the mixed fermentation of fortified milk using strains of *Lactobacillus bulgaricus* and *Streptococcus thermophilus.* The casein (principal protein in milk) is coagulated by the acid produced. A conventional, continuous two-stage process consists of a first-stage, stirred fermenter operating at a pH as low as possible while avoiding casein coagulation. The second stage is a laminar plug flow reactor, in which the acidification is completed and the milk allowed to coagulate under quiescent (inactive) conditions. The pH-Stat continuous fermenter, newly developed, operates during the conventional first stage, in which the feedrate is controlled to maintain a constant pH or end product acid concentration.

Streptococcus thermophilus and *Lactobacilus bulgaricus* are used in mixed culture. The growth medium is a commercial yogurt mix containing approximately 1% milk fat, 12% milk, nonfat solids and 10% sucrose. The growth medium is heat treated at 92°C for 45 minutes and homogenized. The mix is then stored at <5°C.

It has been shown [30] that, separately cultivated, the yogurt bacteria *L. bulgaricus* and *S. thermophilus* ferment the milk rather slowly, but, when combined, their lactic acid production rate is much higher.

Stage of Development

Laboratory-scale research and development is underway.

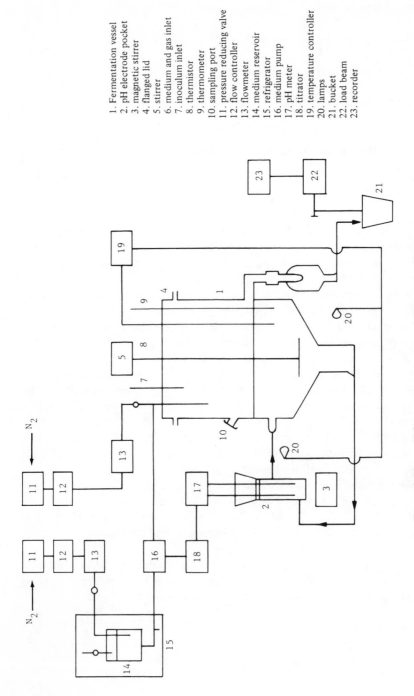

1. Fermentation vessel
2. pH electrode pocket
3. magnetic stirrer
4. flanged lid
5. stirrer
6. medium and gas inlet
7. inoculum inlet
8. thermistor
9. thermometer
10. sampling port
11. pressure reducing valve
12. flow controller
13. flowmeter
14. medium reservoir
15. refrigerator
16. medium pump
17. pH meter
18. titrator
19. temperature controller
20. lamps
21. bucket
22. load beam
23. recorder

Figure 2-23.   pH-Stat continuous-culture fermenter (laboratory scale) [30].

Implications for Energy Consumption

These are not clear at present.

ACIDOPHILUS MILK PRODUCTION BY USING COMBINED CULTURE TECHNIQUE

Description

The conventional means of producing acidophilus milk include mixed incubation of acidophilus cultures and cream cultures with different optimum cultivation temperatures. The disadvantage of such a technique is that the product is sharply acidic and possesses a coarse consistency. This reduces both the dietetic and therapeutic qualities of the product. The process described here allows separate incubation of the two cultures, thus ensuring optimum cultivation of the cultures. To ensure that acidophilus milk, containing bacterial cultures that thrive in dilute acidic milk, exhibits good organoleptic (pertaining to gastrointestinal tract) properties and possesses a high reactivity (containing bacterial cultures that thrive in dilute acidic milk), a new process of manufacture has been introduced. It is based on the separate incubation of milk with both an acidophilus (bacterial) and a cream culture, and their mixing at a ratio of 1:9 after ripening [31].

The process includes two basic incubation steps, which require different temperatures. Of a given volume of pasteurized milk, 90%, with 3.8% fat, is tempered to 30°C and inoculated with 1% cream starter. The incubation is carried out at 23°C for 16-18 hours, resulting in the production of a thick, aromatic coagulum with lactic flavor and acidity ranging from 36 to 40° sulfhydril (SH), where SH is a combined measure of flavor and acidity. The incubating temperature causes the milk proteins to change their structure from normal to random configuration, thus exposing numerous SH groups, which contribute to the lactic flavor. The remaining (10%) pasteurized milk, with a butterfat content of 2%, is heated to 37°C, inoculated with 1% *Lactobacillus acidophilus* starter, and incubated at this temperature for 16-18 hours. The resultant coagulum is thick and has a typical sharply acid flavor and acidity in the range of 85 to 100°SH.

After separate ripening, the coagulum incubated with the acidophilus culture is gently stirred and pumped by a positive pump into the ripening tank, which is provided with a stirrer suitable for intensive but gentle mixing. Here it is mixed with the coagulum of the milk incubated with the cream culture. After thorough stirring, the mixture is cooled to 8-10°C and stored at that temperature until the next day.

Acidophilus milk produced according to this technique has organoleptic properties similar to those of kefir milk, because the ratio of streptococci to lactobacilli (9:1) is about the same in both products. The advantage of this process is that large quantities of acidophilus milk can be produced.

Stage of Development

Commercial production is underway in the United States, Northern Bohemia and Denmark.

Implications for Energy Consumption

None is apparent at present.

CORNSTARCH AS A SUBSTRATE FOR SUCROSE PRODUCTION (LABORATORY)

Description

Commercial production of sucrose from high-fructose corn syrup has been achieved successfully. The process described here and illustrated in Figure 2-24 (the phosphate process) integrates sucrose synthesis with the process of starch digestion and preserves some of the energy of the glycosidic bonds of starch. The only raw material required is starch; the enzymes and phosphate are recycled.

Sucrose synthesis from starch by modification of the conventional process for production of high-fructose syrup includes two phosphorylase enzyme reactions. This process offers several advantages over processes involving sucrose synthesis by condensation of fructose and glucose. The specificity of sucrose phosphorylase precludes formation of other lower-energy products than sucrose. The thermodynamic efficiency of the phosphate process is greater than several of the reversal hydrolysis processes, since some of the energy of the glycosidic bonds of starch is preserved in the glucose-1-phosphate intermediate. The yield of the product sucrose can be enhanced by carrying out the two phosphorylase reactions separately, with manipulation of phosphate concentration.

The disadvantages of the phosphate process include the necessity for separation of glucose-1-phosphate, and the problems associated with insertion of two more enzymatic reactions into the present process for production of high-fructose syrup. The large-scale separation of phosphate and glucose-1-phosphate is an engineering problem.

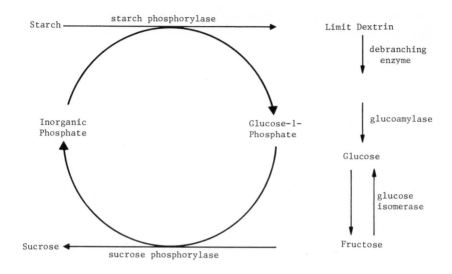

Figure 2-24.   Phosphate process for sucrose production from starch [32].

In the phosphate process, synthesis of sucrose is catalyzed by sucrose phosphorylase from α-D-glucose-1-phosphate and fructose. The fructose is made available from high-fructose syrups, and α-D-glucose-1-phosphate is obtained from starch and inorganic phosphate in a reaction catalyzed by starch or glycogen phosphorylase. The sucrose phosphorylase necessary for the conversion is obtained from *Leuconostoc mesenteroides*. This enzyme is less subject to inhibition by glucose, hence preferred over other preparations. The enzyme is stable after partial purification by ammonium sulfate precipitation. At 4°C it is stable over 49 days and 89% of the original activity is retained. At 30°C the enzyme is stable for 14 days and 32% of the activity is lost over that period. At a pH of 6.52, 10% conversion of fructose to sucrose takes place in 2 hours. Addition of glucose has been shown to lower the rate of depletion of the reducing sugars.

Stage of Development

Laboratory-scale research and development is underway.

Implications for Energy Consumption

Engineering components of the process have not been determined for large-scale sucrose synthesis.

USE OF BACTERIA IN THE PRODUCTION OF FLAVORLESS SOY PROTEIN

Description

Soybean is an excellent source of plant protein from both a nutritional and industrial standpoint. Its application as a food product is very limited, however, because of the characteristic disagreeable flavor. Currently, methods such as steaming, adsorption, washing with solvent, precipitation, and oxidation and reduction are used for deodorization. Use of microorganisms in deodorization is suggested to be a feasible option. The disadvantage of this fermentation approach is that although the process eliminates the offending flavor completely, a secondary fermentation flavor is added to the product. To avoid this problem, a process has been developed utilizing microbial fermentation along with appropriate methods of purification, such as dialysis and adsorption. This process is called the FIC method (Fermentation Isolation Combined). The product known as Bland Soybean Protein Products (BSPP) can be applied to a variety of food products such as dairy products, bakery products and confectionaries, etc.

In the laboratory-scale process, dehulled and defatted soybean flour is suspended in water (ratio 10:1) with a small amount of growth stimulant. The pH of the suspension is adjusted to 6.8 with aqueous KOH, and the suspension is sterilized and cooled to 30°C. The suspension is then seeded with *Lactobacillus brevis* culture and allowed to ferment at 30°C for about 17 hours. During the fermentation, pH is maintained at 6.0 to avoid coagulation of the soy protein. At the end of the fermentation, the broth is adjusted to pH 6.8, sterilized and cooled to below 20°C. Insoluble matter in the fermented broth is separated by sieving and centrifugation, and the fermented broth supernatant is purified by means of adsorption and/or dialysis. The final product is bland soybean protein product dried by spray-drying. Figure 2-25 shows the flowsheet for the preparation of these bland soybean protein products.

Further research is underway to study the physiochemical properties of this product. BSPP has been shown to be applied successfully in milk products, bakery products and confectionaries.

Stage of Development

Laboratory-scale research and development is underway.

Implications for Energy Consumption

They are unknown at present.

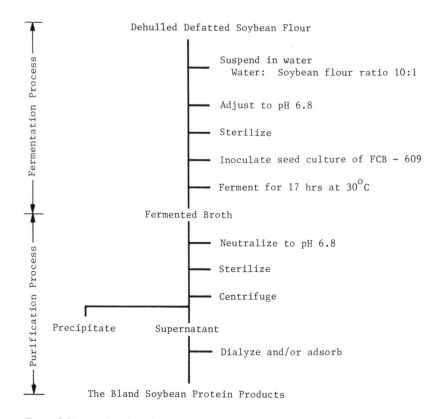

Figure 2-25. A flowsheet for the preparation of bland soybean protein products [33].

ANIMAL WASTE-DERIVED SINGLE-CELL PROTEIN

Description

The uncontrolled release of nitrogen compounds from animal manures can have detrimental effects on aquatic ecosystems and human health. Nitrogen is too economically valuable and too hazardous a material to be released to the environment without an attempt to recover and recycle it in a usable form. Most of the present and proposed methods for handling animal manures result in substantial nitrogen losses and often impose an economic burden on operators. The controlled aerobic fermentation method described here uses the technique of flocculation, which has been shown to recover a high-protein feedstuff, i.e., single-cell protein and residual solids, from the fermenter broth. The fermenter broth consists of bacterial cells and suspended residues from the manure and dissolved solids. These bacterial cells are allowed to flocculate, and the total effluent from the flocculator is sent directly to a centrifuge, thus omitting the step of gravity sedimentation in the traditional waste treatment process. The step of centrifugation permits the operator to recover a high-protein material that can be dried and fed to poultry.

Animal manure and diluted molasses are used as substrate and pumped into the fermenter at a rate that gives a carbon:nitrogen input ratio of ten. Mixed bacteria in animal manure serve as catalysts. The aeration and agitation are maintained at high levels to prevent oxygen limitation. The fermenter temperature is controlled at $25°\pm1°C$, and the pH is maintained at 7.5 ± 0.1 during steady-state operation. The fermentation is allowed to continue for entire residence times. Figure 2-26 shows the experimental apparatus used for the laboratory study.

The process of flocculation uses five inorganic salts—calcium hydroxide, calcium chloride, ferric chloride, ferric sulfate and aluminum sulfate—and three polymeric compounds—chitin, chitosan and Percol 728—in different concentrations and combinations. A flocculant solution containing the above ingredients is added to the samples from the fermenter prior to pH adjustment, and this mixture is agitated at 200 rpm for 3 minutes followed by 3 minutes of additional agitation at 70 rpm. The mixture is allowed to settle for 3 hours and then centrifuged to recover high-protein feedstuff.

Strongly acidic or basic conditions (pH 3 or 12) favor higher percent removal of total solids, as well as a higher rate of flocculation. Calcium hydroxide is shown to be the most effective flocculant; at 20 mg/l and pH 10, calcium hydroxide has been demonstrated to remove 54% of total solids. The high-protein foodstuff recovery from animal manure is suggested to be economically viable.

Stage of Development

Laboratory-scale research and development is underway.

Implications for Energy Consumption

None are apparent at present.

PRODUCTION OF SINGLE-CELL PROTEIN FROM DAIRY WASTE (LABORATORY)

Description

Whey, the by-product of cheese manufacture, creates a worldwide problem of waste disposal of considerable proportion. Whey has been used as an

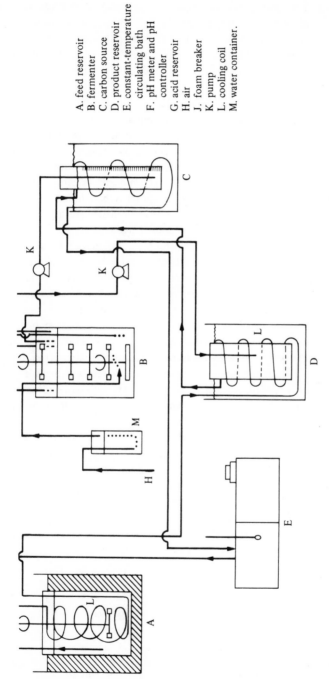

A. feed reservoir
B. fermenter
C. carbon source
D. product reservoir
E. constant-temperature circulating bath
F. pH meter and pH controller
G. acid reservoir
H. air
J. foam breaker
K. pump
L. cooling coil
M. water container.

Figure 2-26. The laboratory process scheme for the single-stage, controlled aerobic conversion of poultry waste into high-protein feedstuff [34].

animal feed and as an additive to cereals. It has also been considered for use in production of SCP and biosynthesis of vitamin B_{12} and vitamin-protein concentrates. The method described here has the potential of using whey in the presence of the mixed yeast-propionibacteria* fermentation as a source of SCP rich in protein and vitamins, while reducing the BOD of the residual fluid.

Propionibacteria and yeast cells are grown in cultures in which stock medium is supplemented with various nutrients. The fermentation of propionibacteria is carried out as follows: equal volumes of 48-hour cultures of *P. shermanii, P. petersonii* and *P. freudenreichii* are mixed, and an inoculum equal to 20% of the volume of whey medium is used. Five liters of inoculum medium are incubated at 30°C for 8 days, and pH is adjusted daily to 6.8–7.2 with NH_4OH. On the sixth day, 5,6-dimethylbenzimidazole (16 mg/l) is added as vitamin B_{12} precursor. The cells are centrifuged at 1900 X g for 1 hour, washed with distilled water and recentrifuged.

The fermentation of yeast-propionibacteria is carried out by using *Kluyveromyces fragilis* since it has been shown to almost completely utilize the lactose in whey medium. A 20-liter amount of yeast-propionibacteria fermentation medium is inoculated with 20% by volume of a 12-hour culture of *K. fragilis* and incubated at 30°C with continuous aeration. After 8–10 hours the lactose content is depleted and the medium is heated to boiling for 5 minutes. Following cooling of the medium, the pH is adjusted to 7.0–7.2, and 16 mg $CaSO_4/l$ is added. A 48-hour culture of the propionic acid bacteria culture (described above) is added at a 1:7 (culture:medium) ratio, and fermentation is continued for 72 hours at 30°C with daily adjustment of pH to 6.8. Four hours before termination of fermentation, 5,6-dimethylbenzinadazole (16 mg/l) is added. The yeast-bacteria cell mixture is spun down at 1900 X g. When the centrifuged cell mass and the culture supernatant are dried together, the protein content is increased by approximately 50%.

The disadvantage of using SCP derived by this procedure for human consumption is that it contains high levels of nucleic acid. The high nuclei acid content of the product (microorganism) mass is a limiting factor in the acceptability of SCP, particularly from bacteria. The metabolic products of the microorganisms, e.g., purines, lead to nucleic acid formation, which, in excessive amounts, can lead to a gout or kidney stone in humans [35].

Stage of Development

Laboratory-scale research and development is underway.

---

*This fermentation medium contains culture of *Kluyveromyces fragilis* and equal volumes of cultures of *P. shermanii l, P. petersonii J.* and *P. freudenreichii J.*

Implications for Energy Consumption.

These are not known at present.

ANIMAL FEED (SINGLE-CELL PROTEIN) FROM PULP AND PAPER WASTE

Description

Pulp and paper waste contains substantial amounts of carbohydrates, which can be used as substrate for the production of protein. The process described here was developed by a Finnish industrial group called SITU, in cooperation with the Finnish government. The process is known as the PEKILO process, in which microfungi are continuously cultivated in spent liquor as submerged cultures.

The PEKILO process utilizes monosaccharides, polysaccharides, carbohydrate derivatives and acetic acid from the spent sulfite liquors. It has been suggested that this process can also utilize waste liquors from the sugar and potato industries.

Figure 2-27 is a simplified diagram of the PEKILO process. The spent liquor is pretreated to remove (strip off) sulfur dioxide. This is performed in a "tray" tower with the aid of steam at atmospheric pressure. The stripped spent liquor is fed to the fermenter in which the pH is regulated with ammonia. The fermentation of spent liquor is a continuous process with a retention time of 4.5–5 hours. Following fermentation, the fungal mycelium (i.e., protein) are separated and washed on a drum filter and thoroughly dewatered. The protein product is dried by utilizing a variety of drying equipment; the main consideration is to preserve the nutritional value of the product.

The fungal protein can be used as an animal feed. Commercial production of single-cell protein is presently being undertaken in Finland. The total power consumption of the full-scale PEKILO process is estimated at 1250 kWh/ton based on an annual production of 10,000 tons.

Stage of Development

There has been commercial development since 1975 at United Paper Mills Ltd. Jamsankoski, Finland.

Implications for Energy Consumption

The total power consumption of the full-scale PEKILO process is estimated at 1250 kWh/ton referred to an annual production of 10,000 tons.

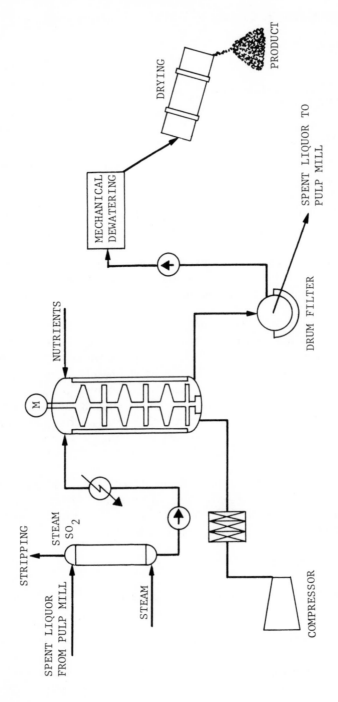

Figure 2-27.   Simplified diagram of the PEKILO process [36].

SINGLE CELL PROTEIN FROM METHANE AND METHANOL

Description

Methane or methanol can be used as a substrate by methane utilizing bacteria to produce bacterial proteins that can be used as supplemental foods or ingredients in animal feeds. Methane-utilizing bacteria *(Methylomonas methanica)* found in water, soils and muds, and marshes are studied extensively. Methylobacteria are grown either by batch or continuous culture on incubating media containing methane (10–98%), oxygen (2–45%), CO_2 (0–20%), sulfates, phosphates and nitrates. The optimum growth temperature ranges from 15–35°C according to strain; the pH range is 5.8–7.4. The oxidation of methane occurs as follows:

$$CH_4 \rightarrow CH_3OH \rightarrow HCHO \rightarrow HCOOH \rightarrow CO_2$$

On an industrial production scale, certain difficulties that can be encountered are: the risk of explosion; the low solubility of methane; the protracted generation time (3–16 hours); and the large quantity of heat released during substrate utilization. Use of mixed cultures of methane-utilizing bacteria improves yield, although maintenance of a mixed culture in equilibrium is difficult over a long period. One potential solution suggested to this problem is the use of chemostat techniques to support stable cultures with bacteria compatible to one another [37].

Methanol, which can be produced readily from methane present in natural gas, is used as the growth substrate for a process developed by ICI (Imperial Chemical Industries), which utilizes *Methylophilus methylotrophus* grown continuously on a sterile medium containing methanol, inorganic salts and ammonia. Figure 2-28 shows the flow diagram of production of bacterial protein from methanol. The pressure cycle fermenter has several unique features: the large quantity of air supplied is used to produce rapid interval circulation and, thus, no stirring mechanism is needed; the height of the fermenter creates a high-pressure region at the bottom where sterile air is introduced, allowing rapid dissolving of oxygen; a low-pressure region at the top causes release of carbon dioxide.

The product contains 72% protein and 8.6% lipids, with an amino acid profile high in lysine and methionine and comparable with that of fish meal. A commercial plant capable of producing 100,000 tons of microbial protein annually is to be constructed at Billingham, England.

Stage of Development

Commercial development is underway.

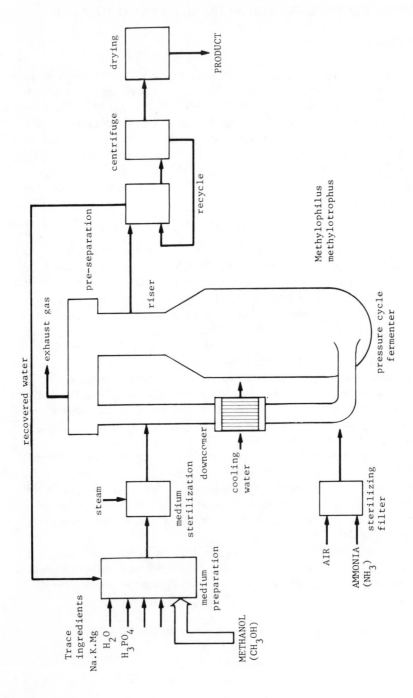

Figure 2-28.   Production of bacterial protein from methanol [38].

Implications for Energy Consumption

None are known at present.

BACTERIAL PROTEIN FROM INDUSTRIAL WASTE GASES

Description

Production of SCP from petroleum or other hydrocarbons is a potentially significant development of this century. The use of lignocellulosic material as a substrate for SCP production also has been studied extensively. A relatively new development is the possibility of using a toxic by-product of the steel industry, carbon monoxide (CO), as a substrate for SCP production [39]. The laboratory process described here has the potential for using different industrial waste gases as substrates to produce protein. The bacteria that utilize CO as their sole source of carbon and energy and are able to grow aerobically are known as "carboxydobacteria." Organisms belonging to this group include *Pseudomonas* sp., *Comamonas compransoris*, *Achromobacter carboxydus* and some unidentified strains. The following reaction is the carbon monoxide oxidation catalyzed by carboxydobacteria in which the CO_2 produced is partially incorporated into cell material:

$$O_2 + 2\ CO \xrightarrow{\text{Carboxydobacteria}} 2\ CO_2 + \text{cell carbon} + \text{energy}$$

An advantage of using these bacteria is that they are resistant to the poisonous impurities contained in most waste gases.

One particular organism, *Pseudomonas carboxydovoraus*, has been studied extensively. This organism grows well in mineral medium under an atmosphere of 50% automobile exhaust and 50% air. The doubling time under these conditions is 20 hours. The carbon monoxide oxidation step is catalyzed by a CO-acceptor oxidoreductase enzyme, which is produced only during growth with CO.

It is suggested that cultivation of carboxydobacteria on carbon monoxide-containing waste gases has the advantages of low cost and ready substrate availability. Lower energy costs can be achieved by cultivating these bacteria under nonsterile conditions.

Stage of Development

Laboratory-scale research and development is underway.

Implications for Energy Consumption

Low energy cost is anticipated.

SINGLE-CELL PROTEIN FROM DAIRY WASTE (COMMERCIAL)

Description

The problems with cheese whey disposal or utilization are numerous. The material is 93.5–94% water, which makes hauling it any distance expensive. Secondly, it is perishable, so cannot be stored any length of time. Finally, treatment processes such as evaporation and drying are capital intensive. A method is described below that offers the commercial conversion of the lactose in whey into high-quality protein, i.e., yeast protein via fermentation. The method also offers a unique closed-loop system producing few, if any, effluents, thus avoiding many of the environmental problems associated with whey treatment and disposal.

During the commercial production of yeast protein, acid and/or sweet, condensed whey is diluted to the appropriate lactose concentration (10–15% whey solids) with water or raw whey. Other additions include phosphoric acid (0.1%), yeast extract (0.13%), ammonia (0.3–0.5%) and hydrochloric acid (0.2–0.5%, pH 4.5). The microorganism used in this process is *Saccharomyces fragilis*. The medium is heated to 80°F (22.2°C) for 45 minutes and then cooled. The fermentation is carried out in a 15,000-gallon (56,755-liter), stainless-steel, deep-tank fermenter that is fully aerated and jacketed. Automatic instrumentation controls used on the system maintain pH (4.5), temperature (90–92°F (32–33.3°C)), aeration (1.0±0.2 volume of air per volume medium), and foaming, as well as levels and volumes in and out of the fermenter. The fermenter is operated in a batch, semicontinuous or continuous manner. Figure 2-29 gives an overall schematic of the fermentation process.

During the whey fermentation, ethyl alcohol is one of the metabolic products formed. By changing conditions of fermentation, better than 90% of the lactose is converted to ethyl alcohol, while the rest is metabolized for cell maintenance, i.e., under anaerobic conditions ethanol production can be increased as shown in Figure 2-30.

The "closed-loop" system designed for this commercial process minimizes waste streams from the fermentation operation, thus reducing potential environmental problems associated with process effluents. Figure 2-31 shows such a closed loop system. In terms of process cost requirements, it has been estimated that a plant capable of an annual production of 4,000–10,000 ton/yr would cost in the range of $6–15 million. With a 1977 dollar value for materials, labor, utility, and investment costs, it is estimated that for a plant with 5,000–10,000 tons annual capacity, total production costs for the feed grade yeast protein would be 11.8–16¢/lb. According to the researchers [40], the fermentation process is an economically viable one.

Stage of Development

Commercial production is underway at Amber Laboratories, Juneau, Wisconsin.

Implications for Energy Consumption

None are known at present.

SINGLE-CELL PROTEIN FROM AGRICULTURAL WASTE

Description

Single-cell protein is of potential value as a supplemental human food. Production of SCP by photosynthetic bacteria is a concept developed to use agricultural by-products and wastes as substrates. Photosynthetic bacteria are abundant in stagnant waters containing decaying organic matter or other organic substances. The growth of photosynthetic bacteria depends generally on the presence of sunlight, the concentration of H_2S and adequate anaerobic conditions. Photosynthetic bacteria is a collective term used for green sulfur bacteria *(Chlorobacteriaceae)*, purple sulfur bacteria *(Thiorhodaceae)* and nonsulfur purple bacteria *(Athiorhodaceae)*.

The basic design parameters of systems that use these bacteria to produce single cell proteins have been extrapolated from experimental data. Factors such as nutrient source, generation time, effect of O_2 on pigment synthesis and photosynthetic growth, pH and optimum temperature are of importance.

The proposed photosynthetic single-cell protein process illustrated in Figure 2-32 has been designed as a semicontinuous process with a mean production capacity of 5 ton/day of SCP. The design is based on the following utilization parameters: (1) a digestor retention time of 24 hours; (2) an operating temperature of 37°C; (3) a pH of 7.0–7.2; and (4) a harvested cellular yield of 10.0 g/l of dried cells. The major processing steps include hydrolysis of the wheat bran slurry, solids removal, neutralization, photosynthetic cultivation, recovery and purification. Major pieces of process equipment and their 1977 costs are itemized in Table 2-12.

In the initial processing step, an aqueous slurry of wheat bran containing approximately 30% solids is acidified with hydrochloric acid (36%). The

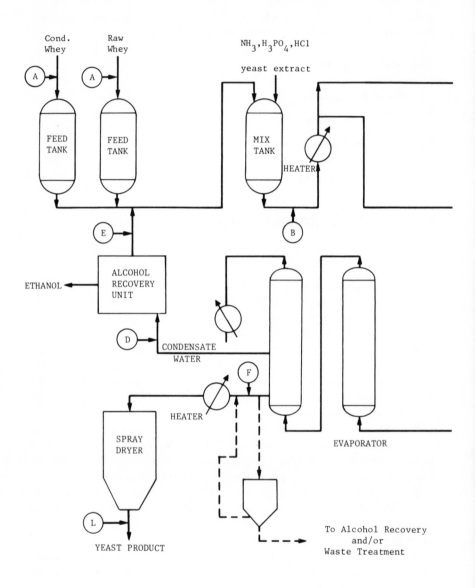

Figure 2-29.   Schematic

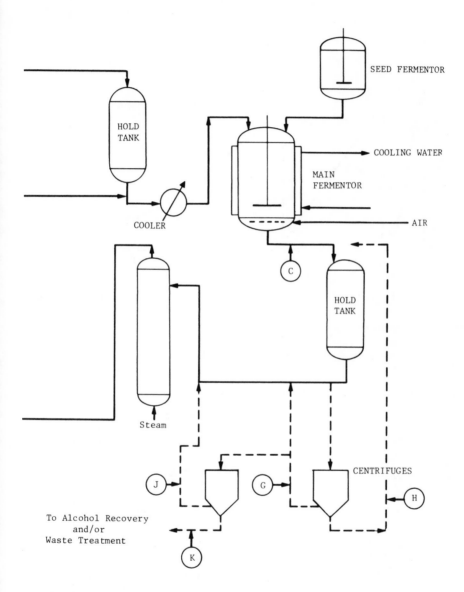

showing monitoring points [40].

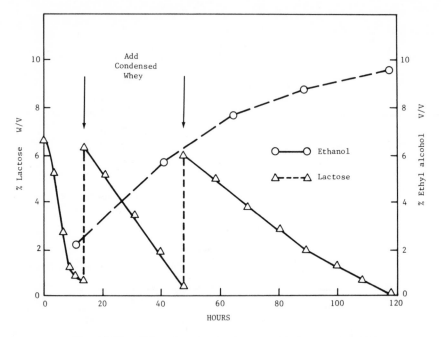

Figure 2-30.   Ethanol production by anaerobic fermentation [40].

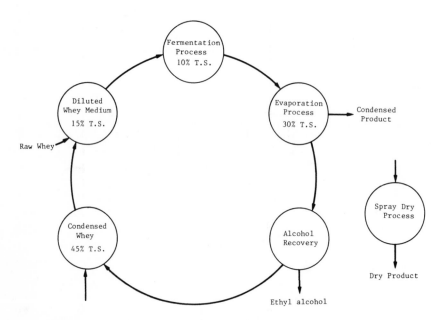

Figure 2-31.   Closed-loop system for fermentation with zero effluent [40].

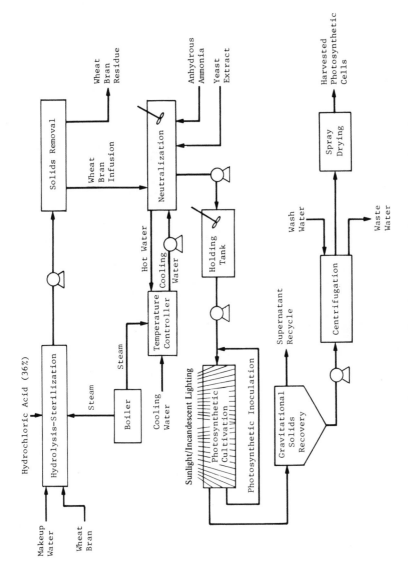

Figure 2-32.   Flowsheet for the photosynthetic single-cell protein process [41].

Table 2-12. Major Process Equipment for Photosynthetic Single-Cell Protein Production [41]

Unit	Number	Description	Unit Cost ($)	Cost ($)
Mixing and Storage Tanks	2	Type: stainless steel; capacity: 400,000 gal	33,000	66,000
Mixing and Storage Tank	1	Type: carbon steel; capacity: 24,000 gal	11,200	11,200
Storage Tank	1	Type: carbon steel; capacity: 12,000 gal	8,400	8,400
Centrifuge	1	Type: solid bowl with conveyor discharge; stainless steel; 90-gpm hydraulic capacity	50,000	50,000
Horizontal Transparent Polyvinyl Chloride (PVC) Pipe	33	Dimensions: 61 cm × 67.1 m; capacity: 4700 gal	22.50/Linear ft.	163,800
Centrifuge	1	Type: disc bowl with nozzle discharge; stainless steel; includes drive and controls; 90-cpm hydraulic capacity	32,400	32,400
Incandescent Lamps	200	400 W; 115 V	6.00	1,200
Boiler Unit	1	Steam capacity; 30,000 lb/hr; steam pressure: 250 psi	52,000	52,000
Pump	1	Type: radial-flow centrifugal; stainless steel; 10 HP max. capacity: 150 gpm	5,380	5,380
Pumps	5	Type: radial-flow centrifugal; carbon steel; 10 HP max. capacity: 150 gpm	1,680	8,400
Spray Dryer	1	Evaporative rate: 4 ton/hr	26,400	26,400
		Total purchased equipment cost:		$425,180

slurry is injected with steam and held at a temperature of 121°C and a steam pressure of 15 psi for 4 hours. The unhydrolyzed wheat bran residue is separated from the supernatant in a solid-bowl, conveyor-type centrifuge. The wheat bran infusion is discharged into a stainless-steel tank in which neutralization with anhydrous ammonia occurs. A yeast extract is added to supplement the infusion medium with vitamins that may have been destroyed during the hydrolysis and heating process. The temperature of the infusion medium is lowered to approximately 40°C at this stage of the process by pumping cooling water from the temperature controller of the photosynthetic cultivation tanks through the cooling jacket of the neutralization tank. Hot water flowing from the cooling jacket is then pumped back to the temperature controller and is used to maintain a temperature of 37°C within the photosynthetic bioreactor. Sterilization of the infusion medium occurs simultaneously during the acid hydrolysis and heating process, and thus the need for an additional medium-sterilization step is eliminated.

The photosynthetic cultivation phase is comprised of illuminating wheat bran infusion pumped from a holding tank at a rate of 500 gph (18,900 liter/hr) into the series of horizontal cultivation tanks. During daytime, direct sunlight is the illuminating source, while at night, 400-volt incandescent lamps lighted by stored solar energy serve the purpose. The wheat bran infusion is retained in the cultivation tanks for 24 hours. The photosynthetic bacteria are inoculated continuously at a rate of 100 gph (378.5 liter/hr) and recovered by pumping spent culture media from the recovery tank to a disc bowl-type centrifuge.

One of the most promising uses of this photosynthetic process is the utilization of photosynthetic SCP for human food supplementation. Photosynthetic bacterial cells contain approximately 65% protein and significant quantities of amino acids such as histidine, isoleucine, leucine, lysine, methionine, phenylalanine, threonine, tryptophan and valine. In addition, photosynthetic cells contain relatively large amounts of ascorbic acid, vitamin D and forms of vitamin B. Unlike other SCP from hydrocarbons in which the possibility of substrate toxicity exists, wheat bran offers the advantage of a readily renewable, toxic-free, natural substrate for the production of edible protein.

Stage of Development

Development is in the pilot plant stage.

Implications for Energy Consumption

They are not known at present.

BACTERIAL PROTEIN FROM POTATO-PROCESSING WASTES

Description

Disposal of large quantities of potato-processing wastes represents a financial burden on the potato-processing industry, an environmental pollution problem and a loss of potential crude protein. The fermentative conversion of hydrolyzed potato-processing wastes by *Lactobacilli* into a livestock feed supplement rich in crude protein is a promising option.

The summarized process involves lactic acid fermentation of acid-hydrolyzed starch obtained from the wash water of potato chip manufacturing. The substrate, the starch cake, also known as potato-processing waste (PPW), is hydrolyzed with concentrated sulfuric acid (0.1–0.7% v/v) by preparing PPW slurry and autoclaving at 121°C for 30 minutes. The hydrolyzed slurry pH is adjusted to 6.8 with NaOH and *Lactobacilli* culture and growth factors are added. The laboratory-scale fermentation process is carried out in a 24-liter fermenter equipped with automatic sterilization, agitation and temperature controls. Figure 2-33 illustrates the basic components of the process. Following approximately 36 hours of fermentation, the efficiency of utilization of reducing sugar is greater than 90%. The concentrated product is a light brown, syrup-like material, resistant to microbial spoilage. Here, lactic acid that is produced serves as a trapping agent for ammonium ion. The trapping of ammonium ion greatly increases the crude protein level in the product, thus increasing its value as a feed supplement. This process also has a potential of additional protein production for human consumption.

The advantages of using *Lactobacilli* in fermentation processes are as follows:

1. *Lactobacillus* sp. grow well at the relatively low pH of 5.5 and at the relatively high temperature of 43°C. These conditions are unfavorable for the growth of most contaminant microorganisms.

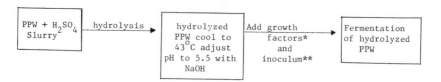

* Growth factors include minerals, yeast extract, trypticase and CO_2
** Inoculum – <u>Lactobacillus</u> <u>acidophilus</u>, <u>Lactobacilius</u> <u>thermophilus</u> and <u>Lactobacillus</u> <u>bulgaricus</u>.

Figure 2-33.   Schematic of the potato-processing wastes fermentation [42].

2. *Lactobacillus* sp. are homofermentive, producing lactic acid as the major product of their metabolism, thus allowing efficient conversion of substrate to product.

3. *Lactobacillus* sp. have amino acid profiles better than FAO reference protein.

Stage of Development

Laboratory-scale research and development is underway.

Implications for Energy Consumption

They are unknown at present.

BACTERIAL PROTEIN FROM URBAN SOLID WASTE

Description

A protein from cellulose process developed at Louisiana State University (LSU) uses a unique microorganism of the genus *Cellulomonas* to increase the speed of cellulose degradation. This process, based on a laboratory-scale design, is expected to increase single cell protein productivity by a factor of 10, from the conventional 0.1 g/l/hr to 1.0 g. This upgrading has made the process economical, and the developer anticipates four to six years of advanced development leading to a large-scale demonstration plant and full commercialization [43].

The process has two main stages. First, chopped cellulosic seed material, such as crop wastes or separated urban solid waste, is treated with a strong alkali, such as sodium hydroxide, in solution with a catalyst such as cobalt (II) chloride. The separated, treated cellulosic material is then heated to moderate temperature (25°–100°C). This step breaks down the tough, protective coating of lignin over the cellulosic fibrils and destroys much of the crystalline structure of these fibrils. The treatment also eliminates contaminants in the feed material.

After this digestion step, treated cellulose cake is neutralized with acid and passed at a pH of 5-9 into an aerobic fermentation chamber. Here the cellulose cake is attacked at moderate temperatures (25°–40°C) by the *Cellulomonas* bacteria. The bacterial attack actually proceeds by means of two enzymes produced by the bacteria. One of these enzymes breaks the linkage between maple sugar units in the cellulosic chain to form mainly disaccharides. The second enzyme attacks the disaccharides and renders them digestible by the microorganisms.

During fermentation, the bacteria consume about half of the cellulose while the remaining cellulose is recycled. In other words, the process gets a 50% yield of cell mass from the cellulosic content of waste fed into the process. This cell mass is about 60% protein. After a series of separation and drying steps, the final product emerges as a dry, granular material.

The protein derived by the LSU process exhibits a favorable combination of amino acids of a quality between vegetable and animal protein. However, some nucleic acids accumulate with the product protein, blocking human use of the product. Research is underway to find means to remove these nucleic acids.

Since the publication of the first report, the LSU scientists have been able to develop a continuous laboratory process. Using a 40,000-liter fermenter and a reaction time of 20 hours, they were able to utilize 80% cellulose to produce protein and 60% of the cellulose was converted to glucose [44].

Stage of Development

Laboratory-scale research and development is underway.

Implications for Energy Consumption

There is good potential for reduction in energy requirements and cost.

CHAPTER 3

ENERGY

This chapter addresses those applications directly involved in the production of fuels such as alcohols, methane, hydrogen, etc.

CELLULOSIC WASTE TREATMENT TO PRODUCE ALCOHOL

Description

Enzymatic conversion of cellulosic waste to glucose with various applications as food source, feedstock, fuel and supplemental food can be
achieved economically. The enzymatic conversion has added advantage
over acid hydrolysis in terms of enzyme specificity, nonreactibility with
impurities, and purity of end product.

A process developed by the U.S. Army Natick Research and Development
Command is based on the use of cellulase derived from mutant strains of
the fungus *Trichoderma viride*. The schematic diagram of this process is
shown in Figure 3-1. The production of enzyme is accomplished by growing
the fungus *Trichoderma viride* in a culture medium containing shredded
cellulose and various nutrient salts. Following its growth, the fungus culture is
filtered. The clear straw-colored filtrate solution is used in the saccharification reactor. This step proceeds by a step in which cellulase activity of the
filtrate is measured and acidity adjusted to a pH of 4.8. Milled cellulose
is then introduced into the enzyme filtrate solution and allowed to react
with the cellulase to produce glucose. Saccharification takes place at
atmospheric pressures and at a temperature of 50°C. The unreacted cellulose
and cellulase are recycled back into the reactor vessel. The crude glucose
syrup, which is an intermediate product, is filtered for use in microbial
fermentation processes to produce chemical feedstocks, single-cell proteins,
fuels, solvents, etc.

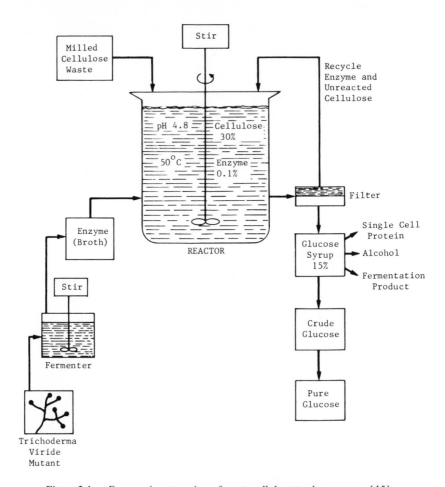

Figure 3-1.   Enzymatic conversion of waste cellulose to glucose sugar [45].

To make this process economically viable, further research is needed in the following areas [45]:

1.   production of high-quality cellulase enzyme from *T. viride* capable of hydrolyzing insoluble crystalline cellulose;
2.   optimization of the hydrolysis reaction to achieve higher sugar production rates;
3.   development of cost-effective pretreatment for cellulosic materials to make them readily susceptible to enzymatic hydrolysis; and
4.   development of adequate cellulosic feedstock.

Figure 3-2 illustrates a pilot-plant process for cellulose production currently in use at Natick R&D Laboratories. According to the researchers, through manipulation of variables such as agitation speed, dissolved oxygen, nutrients, fermentation temperature and acidity of the fermentation solution it has been possible to double and, at some times, quadruple the enzyme productivity of the process in the pilot plant. Both batch and continuous enzyme production techniques are being investigated.

Parameters to be optimized for the most economical production of fermentable sugars during the hydrolysis process include residence time, enzyme activity, operating temperatures, acidity of reacting slurry, susceptibility of cellulosic feedstock to enzymatic hydrolysis, concentration

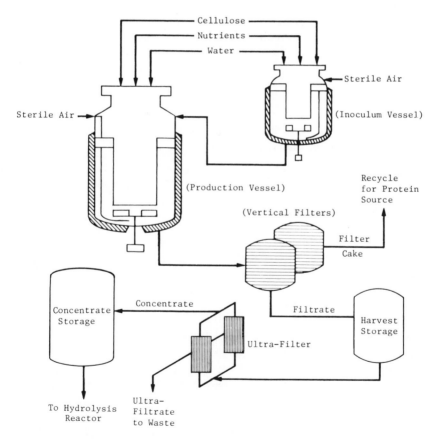

Figure 3-2.   Pilot-plant process for cellulase production [45].

of inerts in feedstock, degree of agitation and mixing of the reacting slurry, and concentration of syrups in the reaction vessel. The pilot-plant process for hydrolysis for which the above parameters are being optimized is shown in Figure 3-3.

Wastes recovered from urban refuse, industrial wastes, agricultural residues and feedlot wastes have been used as substrate for the pilot plant process. Research includes the pretreatment of cellulosic waste for hydrolysis. The process described here is technically feasible, and planning is under-way for the design of demonstration plants [45].

The crude sugar syrup, which is an end product of the saccharification step, can be used to make ethanol. Currently, industrial ethanol is produced from ethylene by two principal processes given in Figure 3-3.

The fermentation ethanol is produced by the anaerobic growth of yeast on fermentable sugars obtained from cornstarch or other agricultural products or residues. This technique also can be applied to crude sugar syrup. The ethanol fermentation is a two-step process in which first the cellulosic material is hydrolyzed to produce fermentable sugars. These sugars are used, in turn, to produce ethanol by yeast fermentation.

Stage of Development

The prepilot-plant stage involves hydrolysis of cellulosic waste. Ethanol fermentation is carried out at commercial plants.

Implications for Energy Consumption

These are not clear at present.

METHANE PRODUCTION USING DAIRY MANURE (DIGESTOR OPERATION)

Description

Anaerobic digestion of organic material to produce energy is a well-studied concept. This concept was used in development of a full-scale anaerobic digester built on the Washington State Dairy Farm in Monroe, Washington. The digestor has been in operation since mid-1977. The gas produced is to be used to fuel the boiler in the dairy farm's creamery. Figure 3-4 is a schematic of the Monroe anaerobic digestion system. This particular Washington State dairy farm has 250 acres of farm area with 400 head of Holstein cattle; a milking herd that varies from 180 to 200 cows; and a

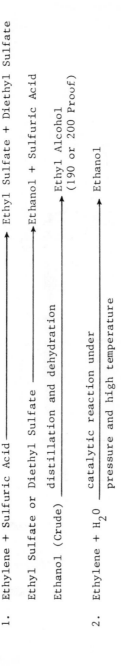

1. Ethylene + Sulfuric Acid $\longrightarrow$ Ethyl Sulfate + Diethyl Sulfate

Ethyl Sulfate or Diethyl Sulfate $\xrightarrow{\text{distillation and dehydration}}$ Ethanol + Sulfuric Acid

Ethanol (Crude) $\longrightarrow$ Ethyl Alcohol (190 or 200 Proof)

2. Ethylene + H$_2$O $\xrightarrow{\text{catalytic reaction under pressure and high temperature}}$ Ethanol

Figure 3-3.   Equations depicting conversion of ethylene to industrial ethanol.

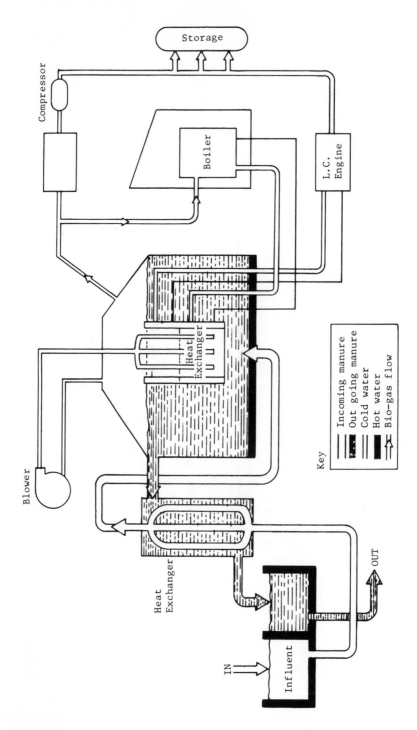

Figure 3-4.   Schematic of Monroe anaerobic digestion system [46].

creamery to process milk, cottage cheese and ice cream for use in government institutions. The manure from these dairy animals is used in the digestor.

The digestor design is a modified version of a municipal sewage treatment plant digestor. It consists of the digestion tanks, the manure handling system, the digestor heating and mixing system, and the gas handling and utilization system. The digestor tanks are two 189-m^3 reactors fitted with silo roofs. The fixed cover, glass-lined (corrosion-resistant) tanks are 7.82 meters in diameter and 4.57 meters in height. The tanks are insulated with styrofoam to avoid freezing during winter. The digestor operation is simplified by using gravity flow rather than a pump to move effluents from the digestor to the storage lagoon. This reduces the energy consumption by eliminating pump requirements.

The digestor operated at 35°C is equipped with a gas recirculation mixer, which requires considerable electrical energy and regular maintenance. The use of the mixer is reduced by allowing mixing to occur naturally in the tank due to natural convection currents and gas bubbling. Elimination of gas by recirculation mixing does not affect the total gas production. The electrical energy savings have been demonstrated to be about 60 GJ per month, which represents about 90% of the original electrical demand of the system and a significant portion of the net energy yield. This, in turn, offers savings in capital, operating and maintenance costs.

The manure handling poses a difficult problem in operating a full-scale digestor. The centrifugal pump has been proven to be the most reliable for dairy manure slurries, containing less than 10% total solids. For thicker slurries, shorter pipes and fewer fittings are recommended.

The gas handling system is designed to burn the gas to produce process steam in the farm creamery. The steam produced can also be used to produce electricity in emergency situations.

Startup of the dairy manure digestor is relatively easy. Failure, in general, is due to lack of seed material or buffering capability. The digestor system must be shut down occasionally to remove or skim off sediment that reduces efficiency.

Table 3-1 gives the amount of gas produced by this digestor system over 24 months of operation. It has been demonstrated that gas production varied approximately linearly with the manure loading rate, while there was lesser variation in microbiological efficiency. The overall gas production rate average is shown to be about 178 m^3/day. This gas is used as a fuel for a boiler for digestor heating, emergency electricity generation, and heating and cooling in the lab; 47% of the total gas produced is consumed for these purposes.

The potential annual total energy production of the system is about 1800 GJ. The largest energy consumer in this digestor system is the digestion tank maintained at 35°C. Figure 3-5 illustrates the electricity consumption

Table 3-1. Gas Production and Boiler Consumption [46]

Month	Gas Production (m³/day)	Daily Load (kg vs/m³)	Boiler Consumption (m³/day)	Percent Total Gas Production Consumption by Boiler
October[a] 1977	93	3.09	62	67
November[a]	107	3.68	75	70
December	147	4.72	64	44
January 1978	146	4.85	52	36
February	198	6.63	79	40
March	193	5.99	88	46
April	216	6.11	91	42
May	221	6.51	88	40
June	241	6.45	73	30
July	201	4.84	58	29
August[a]	108	2.66	37	34
September	183	5.13	74	40
October[a]	144	3.35	68	47
November[a]	145	3.53	92	63
December[a]	154	3.90	90	58
January[a] 1979	108	2.90	85	79
February	206	5.54	98	48
March	234	6.42	104	94
April	230	5.79	80	35
May	245	6.28	77	31
June	218	5.33	68	31
July	224	6.97	70	31
August[a]	132	3.23	44	33

[a]Low production during these months was due to:

1. October/November 1977–the startup procedure of loading low solids of 4–8% TS.
2. August 1978–an 11-day outage of the loading pump for major repairs.
3. October–December 1978–low loading rates resulting from incomplete scraping.
4. January 1979–a 15-day period of freezing temperatures and no scraping.
5. August 1979–a 14-day digestor transfer period of no loading.

of the system over a 22-month period. This reduction in consumption is achieved by natural mixing of manure, transporting slurry by gravity flow, and reducing water content of the slurry. This digestor system is not yet at optimum energy reduction capacity [46].

Stage of Development

Full-scale operation in Monroe, Washington.

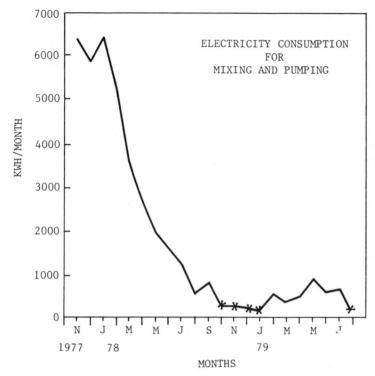

* Electrical Consumption low due to low loading rates.

Figure 3-5. Electricity consumption [46].

Implications for Energy Consumption

Energy-efficient process. Total energy use reduction from 6500 kWM/month to less than 500 kWM/month over a 22-month period.

PRODUCTION OF HYDROCARBONS (LIQUID) FROM ALGAE

Description

The concept presented here is of commercially converting algae into liquid hydrocarbon fuels and chemical feedstocks by reacting the microscopic green plants (algae) with expensive (i.e., commercially obtained) hydrogen.

In the laboratory-scale process, the liquid obtained is similar to a light gas oil. The slightly brown oil is composed of hydrocarbons that are 18-24 carbon atoms long. Further work is needed to characterize the product. This process simultaneously produces ammonium carbonate along with light gas oil that can be decomposed to ammonia and carbon dioxide, which are needed by the growing algae. This process, which involves reacting hydrocarbons with steam over a catalyst, requires an external input of energy. This energy potentially could be provided by recycling the product hydrocarbons [47].

In the batch experiment, 20-25 g dried algae is treated with hydrogen. First, the dried algae is mixed with mineral oil and the resulting slurry fed into an autoclave along with hydrogen at 1000 psi pressure and a cobalt molybdate catalyst. The algae is then heated under these conditions for about one hour at 400°C. This results in an initial breakdown of the algae into a tar-like intermediate called asphaltene. Asphaltene is then gradually hydrogenated into hydrocarbon liquids; gases such as methane, ethylene, ethane, and ammonium carbonate; and water.

To make this process cost-effective it probably has to be operated continuously and, preferably, in sunny, warm regions [47]. It is estimated that a 200-m^2 algae farm could provide approximately 30,000 billion barrels per day to supply a medium-sized refinery. The step of drying the algae is an energy-intensive step; however, part of this energy can be supplied by diverting some of the reaction products to serve as a fuel or feed for the reactor. It is envisioned that some hydrogen and fertilizer would have to be supplied from outside the so-called "closed-loop" system.

Further research is needed to estimate costs and energy requirements. It should be noted that this work has been criticized by others. In their viewpoint, the process is costly and the energy required will be more than the energy (oil) recovered from the process [48]. It is anticipated that the process will not be economical due to the large area requirements and the high cost of drying.

Stage of Development

Laboratory-scale research and development is underway.

Implications for Energy Consumption

Drying the algae is an energy-intensive step. Product hydrocarbons (light oil) can be recycled, however, to provide energy for the drying step.

IMPROVED ETHANOL FERMENTATION THROUGH
INCREASED FERMENTER PRODUCTIVITY

Description

During conventional ethanol production by fermentation, the major constraint is end-product inhibition. In other words, when concentrated sugar solution is fermented and the ethanol concentration of the fermentation broth increases above 7–10%, the ethanol concentration inhibits yeast growth and, in turn, suppresses the ethanol production rate. Currently, to maintain the ethanol concentration below inhibitory levels, the concentrated sugar solution is diluted to 10–20%. This results in lower fermentation rates per unit volume, lower yeast growth and use of larger fermenters, mixing and storage tanks, heat exchangers and distillation columns. To avoid this problem of ethanol inhibition and resultant higher capital and operating costs, an alternative process has been suggested. In this laboratory-scale process, the high volatility of ethanol is used to advantage by boiling off ethanol by vacuum operation, thus increasing fermenter productivity.

In this procedure, the organism used is *Saccharomyces cerevisiae*. The medium is steam sterilized at 121°C for 30 minutes and cooled to ambient temperature. The glucose solution and mineral solution are sterilized separately to avoid caramelization of the glucose and are mixed following cooling.

The vacuum fermenter is the 5-liter "Mino Ferm" fermenter. A schematic diagram of the vacuum fermentation system is given in Figure 3-6.

The only modification in the system is addition of a 1500-watt heater constructed of four 10-inch (254 mm)-diameter coils of 0.5-inch (12.7 mm) copper tubing wrapped with electrical heating tape. The heat input is controlled by variable autotransformers. The vapor generated in the fermenter is condensed on the shell side of the heat exchangers by a 10% methanol-water solution chilled to −4°C. The condensate is collected in a 40-liter stainless steel receiving tank set in a dry ice bath. The vacuum system is run by a vacuum pump. The liquid level control and vacuum control allow adjustment of the feedrate of the fresh fermentation medium and ethanol boiloff rate, respectively.

The vacuum fermenter system is sterilized by boiling a 70% ethanol-water solution for 8 hours and then flushing the system with air for 4 hours to remove the last traces of the sterilizing solution. The fermenter is filled with 3 liters of 10% glucose medium at 35°C and inoculated. Following 12-16 hours of fermentation, the pressure in the fermenter is decreased slowly (25 mm Hg/min) until the fermentation broth begins boiling at 35°C. The ethanol boiling is maintained at 50 mm Hg/min pressure. At

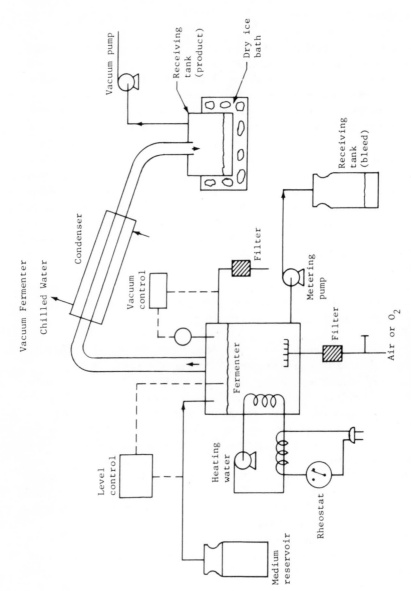

Figure 3-6.   Schematic diagram of the complete vacuum system [49].

this pressure the boiling point of the fermentation broth containing 1% ethanol is 35°C–the optimum fermentation temperature of the yeast. The pH is maintained at between 3.5 and 4.0, and oxygen is provided at a rate of 240 ml/min at STP (standard temperature and pressure) and an agitation rate of 500 rpm.

The *Sacchromyces* cells are recycled by use of the settler arrangement shown in Figure 3-7. The cell recycle technique facilitates an increase in cell mass concentration, which, in turn, produces higher fermentation rates per unit volume.

This system has been tested under semicontinuous and continuous vacuum operations. During semicontinuous operation, the cell concentration and ethanol productivity steadily increase with time over 48 hours. However, after 48 hours both decline, which may be due to the accumulation of nonvolatile components in the fermenter resulting in a yeast kill. During long-term continuous operation, constant withdrawal of fermented broth

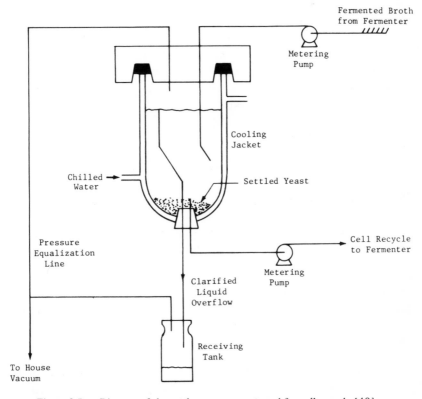

Figure 3-7.   Diagram of the settler arrangement used for cell recycle [49].

maintains a concentration factor of 7.7. This operation allows production of 40 g/l-hr ethanol. It has been noted that with conventional continuous fermentation at STP with the same organism and similar fermentation media, 7 g/l-hr can be obtained. Thus, the vacuum system produces an almost sixfold increase in ethanol productivity compared to a conventional continuous operation.

The use of the settler-vacuum system provides almost a twelvefold increase in productivity over that obtained in conventional continuous operation. In an industrial application of such a system, the settler would be substituted by a centrifuge. This would reduce operating and capital costs.

The following summarizes the advantages and distinctive features of the vacuum technique for ethanol fermentation. The main advantage of the vacuum fermenter is the elimination of ethanol inhibition. This permits concentrated sugar solutions (33.4%) to be fermented at fast rates. The cell recycle in the vacuum system increases the ethanol productivity to a level almost twelve times that obtained with conventional fermentation. The direct consequence of this increased productivity is suggested to be a twelvefold reduction in fermenter volume required for an industrial ethanol fermentation. This reduction in fermenter volume could result in reduced energy consumption. Further studies are needed to determine the extent of impact on energy use. It has been suggested that fermentation-derived ethanol may some day serve as a supplement to, or even a replacement for, conventional petroleum liquid fuels.

Stage of Development

Laboratory-scale research and development is underway.

Implications for Energy Consumption

There is a higher yield of ethanol over a reduced time period, thus there is the potential for a reduction in energy use.

METHANE FROM DAIRY MANURE (PLUG FLOW REACTOR OPERATION)

Description

Anaerobic digestion technology has been used in the sewage sludge digestion. Currently, this fermentation technique is being evaluated for energy production from a range of waste organics and agricultural by-products. Choosing appropriate design criteria for different organic substrates such as domestic sewage, food processing wastewater, sewage sludge, dairy manure, municipal solid waste, poultry manure, wood and straw is critical. The constraints that these substrates impose on the reactor design include the dilute nature of the substrate, its availability in composition and quantity and its low temperature. Figure 3-8 illustrates the steps in the conventional processing of biomass and the new, optimized design concept outlined here. The new design concept eliminates many of the treatment and handling steps, thus making it economically feasible.

The two full-scale, dairy system methane generators that are under operation over the last two years are designed based on results from laboratory and pilot-plant operations. The reactor known as a plug flow reactor consists of soil-supported structures lined and covered with flexible rubber-type material. The reactor is fitted with temperature controls and pumps and the entire system is insulated from the top. Figure 3-9 is a schematic of the full-scale 65-cow dairy plug flow reactor.

Following two years of full-scale plug flow reactor operation, the researchers have performed a cost analysis (Table 3-2). For all farm sizes considered, the cost of energy production is less than existing conventional sources.

Stage of Development

There are full-scale reactors at Cornell University, New York.

Implications for Energy Consumption

This process represents a self-sufficient energy system with relatively low capital and maintenance costs.

Table 3-2. Cost Analysis Results [52]

	Dairy Reactor Capacity, Number of Cows		
	25	100	500
Net Energy (GJ/yr)	203	916	4,920
Fermenter Cost ($)	11,900	20,800	53,500
Actual Net Energy Cost ($/GJ)	$6.73	$3.07	$1.89
Payback Period (yr)	6.3	2.9	1.7

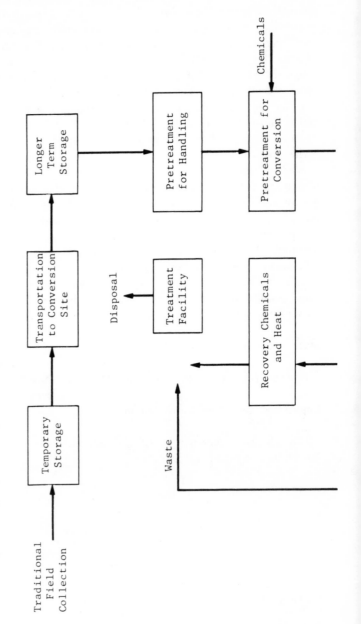

A. Conventional System for Dryer Biomass

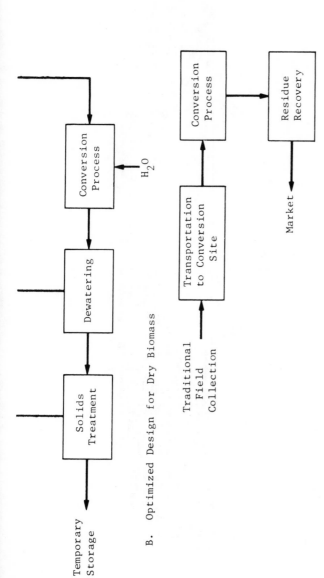

Figure 3-8.   Comparison of steps in processing of field-dried biomass in conventional and a new optimized design concept [50].

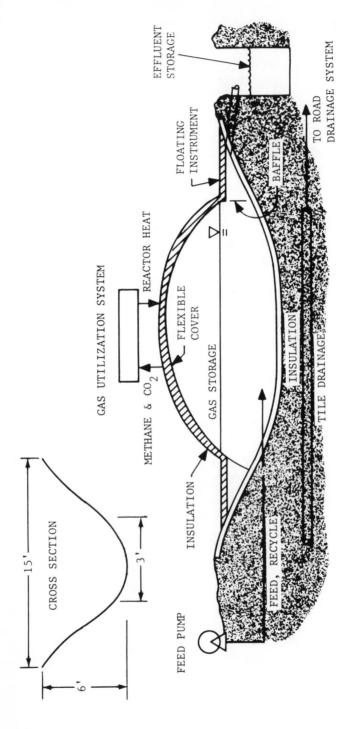

Figure 3-9.   Schematic of the full-scale plug flow reactor [51].

METHANE GAS PRODUCTION FROM PLANT MATTER

Description

The use of agricultural and forest crops and their residues for fuel is not a new concept. Plant matter is an easily usable and renewable fuel. In the process described here, plant matter is used as a substrate for anaerobic digestion to produce synthetic natural gas. Methane generation by anaerobic digestion can be carried out with mixed cultures of microorganisms; sterile feed and pure culture inoculations are not required. The basic requirement of such a process is that the digestor feed must contain sufficient nitrogen and phosphorus salts, along with trace nutrients, for maximum activity and good growth of microbes. Another important requirement in terms of design and operation of an anaerobic digestion system is the retention time. The greater the retention time, the greater the fraction of the organic matter in the feed that is digested.

Woody species such as sycamore, soft maples, hybrid poplars and cotton woods are well suited for anaerobic digestion due to their relatively low lignin content and relatively high hemicellulose contents. Woody plant matter has been shown to produce equal amounts of methane and carbon. This methane can be tapped as an energy source.

Pretreatment of woody plant matter is necessary prior to the digestion. The proposed pretreatments include use of strong acid or alkali, exposure to SO_2 and NH_3, chemical pulping, irradiation with high-energy electrons, steeping in steam or hot water, grinding into fine particles and enzyme treatments. For a feasible, economic process for gas production, steeping in steam or hot water combined with grinding is suggested to be the most suitable.

Here, green deciduous wood chips are first fed by means of a rotary valve into a double-revolving-disc attrition mill under pressurized conditions. It is estimated that 1.21 MJ/kg of energy will be required to grind the chips to about forty mesh size. This material is then immediately dropped down into a steeping tank where ground wood particles are dispersed to obtain feed slurry to be delivered to the anaerobic digestors. The water needed in this step is made available from filtrate separated from the digestor effluent. Solubilization occurs in the steeping tank at 190°C. The partially solubilized slurry is cooled and then fed to the pH-adjustment tank (pH slightly above neutral), then pumped directly to the anaerobic digestors. After retention time of 15 days, the spent slurry is vacuum filtered.

The gas evolved from the anaerobic digestors is a mixture of methane and carbon dioxide saturated with water vapor at 60°C and at a pressure of only a few inches of water above atmospheric pressure. This gas must be compressed to 1000 psi, have CO_2 removed, and be dried to make it pipeline quality. The gas purification is achieved by first compressing to 300 psia, removing the carbon dioxide and then compressing the remaining methane to the required pressure. This is followed by drying in a glycol dehydration unit in which the moist methane is contacted with thriethylene glycol in an absorber. Figure 3-10 shows the schematics of the anaerobic digestion process.

It has been suggested [53] that the required investment, based on 1974 prices, of about $3.00 U.S. per daily standard cubic feet (SCF) of capacity ($106/m^3/day) for an SNG plant using this process is comparable to the estimated cost for a coal gasification plant. The operating and maintenance costs can be reduced by the proposed improvements in the proposed process:

- a grinding energy requirement reduced to 0.99 MJ/kg rather than 1.21 MJ/kg;
- reduction in retention time in the digestors by three days;
- increased solubilization of plant material in the steeping tank; and
- a methane:carbon dioxide ratio in the digestor offgas of 60:40, rather than 50:50.

These improvements could reduce the investment per daily SCF by about $1.00 to less than $2.00 ($71/m^3/day).

Stage of Development

Laboratory-scale research and development is underway.

Implications for Energy Consumption

It represents an alternative means for natural gas production.

IMPROVED OIL RECOVERY USING MICROBIAL LEACHING TECHNIQUE

Description

Present technology for production of oil from shale centers around a process of retorting. Retorting technology has some limitations listed below [54]:

1. Retorting requires an investment of thermal energy that must be subtracted from total energy production.
2. Retorting is inefficient, releasing about 50% of available hydrocarbons.
3. Formation of high-molecular-weight aromatic compounds inhibits refining efficiency, and they have been found to be carcinogenic.
4. Retorting results in large volumes of spent shale, which presents a disposal problem.

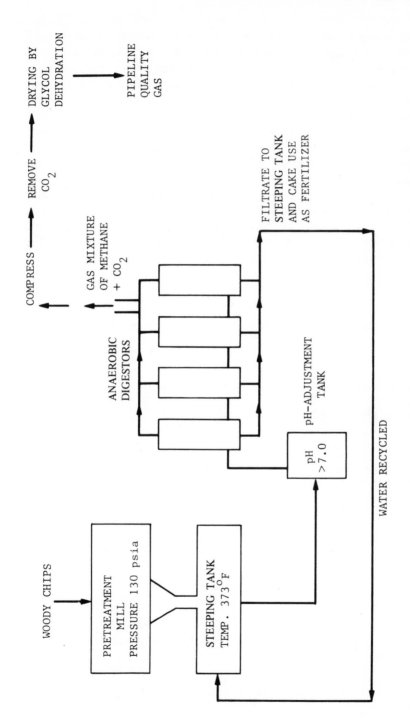

Figure 3-10.   Schematic of anaerobic digestion process to produce synthetic natural gas (SNG).

Bioleaching is an alternative shale technology that is currently under laboratory investigation. During bioleaching, sulfur-oxidizing bacterial species of the genus *Thiobacillus* (capable of producing sulfuric acid) are used. *Thiobacillus* in the presence of oil shale can use portions of the inorganic mineral matrix as a nutrient source. The acid medium produced by *Thiobacillus* has been shown to attack and chemically degrade the carbonate portion of the kerogen-entrapping inorganic mineral matrix. This develops porosity and permeability and increases the availability of internal surface in contact with the leaching medium. Table 3-3 illustrates the properties of bioleached oil shale.

A continuous cyclic process has been developed that uses the bioleaching principle outlined above. Raw shale is continuously leached by gravity with a culture of *Thiobacillus thiooxidans*. The leachate is collected and supplemented with nutrients for growth of the sulfate reducer. The leached medium is inoculated with another autotrophic microbe, *Desulfo-*

Table 3-3. Properties of Bioleached Oil Shale, Laboratory Scale [54]

Properties	Efficiency
Total Weight Loss	35–40%
Density Change	From 1.98 reduced to 1.21 cm^3/g
Enrichment of Organics	From 15 increased to 28%
Carbonate C Analysis	85–93% removed
Dolomite by X-ray	Peak disappeared
Pore Volume Opened	Nil to 50–100 μ^3
Oxidation Rate	3- to 4-fold increased
Oil Yield	Ca 30% increased
Retorting Temperature	At least 50°C lower

Table 3-4. Schematics of a Cyclic Method [54]

Thiobacillus spp.	Sulfur oxidizer	$S + 1\frac{1}{2}O_2 + H_2SO_4 \rightarrow H_2SO_4$
	Dissolution	$CaCO_3 + H_2SO_4 \rightarrow CO_2 + H_2O + CaSO_4$
	Cell synthesis	$CO_2 + H_2O \rightarrow CH_2O + O_2$
Desulfovibro spp.	Sulfate reducer	$CaSO_4 \rightarrow CaS + 2O_2$
	Utilization of dead cell	$CaSO_4 + 2(CH_2O) \rightarrow CaS + 2H_2O + 2CO_2$
	Hydrolysis	$CaS + 2H_2O \rightarrow H_2S + Ca(OH)_2$
	Recycle of sulfur	$H_2S + \frac{1}{2}O_2 \rightarrow S + H_2O$

vibro vulgaris, and anaerobic conditions are initiated by a pyrogallic acid seal. Growth is detected by the characteristic blackening due to FeS formation, odor from H_2S and a decrease in sulfate ion concentration. Following 60 hours of inoculation, sulfate ion concentration is reduced from 1.6 to 0.8 g/l. The schematics of a cyclic method are given in Table 3-4. Figures 3-11 and 3-12 are flow diagrams of the bioleaching process and

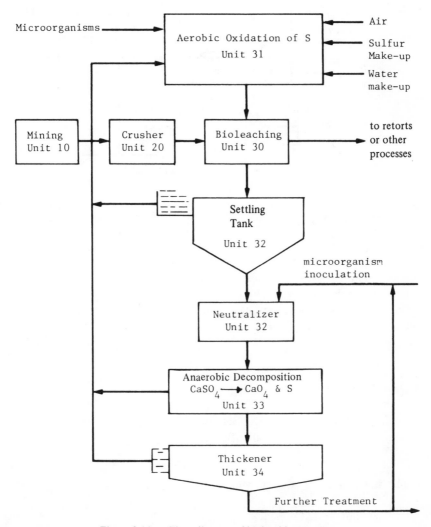

Figure 3-11. Flow diagram of bioleaching process.

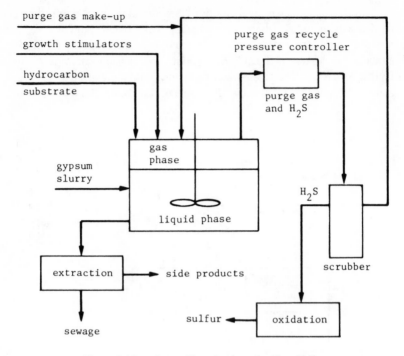

Figure 3-12.   Anaerobic reduction of sulfate [54].

the anaerobic reduction of sulfate, respectively. Figure 3-13 is a conceptual diagram for in situ bioleaching.

The process described here has a long-range goal of changing a shale-bearing basin into an oil- or gas-producing basin that can be recoverable by current technology.

Stage of Development

Laboratory-scale research and development: the work has been discontinued due to some technical-difficulties [55].

Implications for Energy Consumption

These are not clear at present.

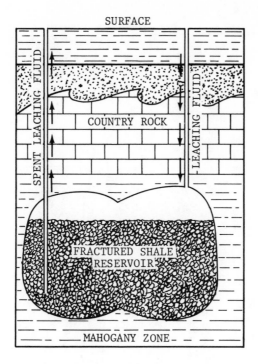

Figure 3-13.   In situ bioleaching [54].

UTILIZING URBAN WASTE FOR THE PRODUCTION OF METHANE GAS

Description

A substantial body of knowledge exists related to the microbial production of methane from organic materials. This technology is being applied in a demonstration plant for producing methane from urban refuse in Pompano Beach, Florida. This summary is based on 1977 information.

The process flow diagram is given in Figure 3-14. With existing technologies developed for producing refuse-derived fuel (RDF), it is possible to produce a light fraction organic that can be used as substrate for the anaerobic fermentation process. Refuse is passed through a shredder, then through a trommel screen to remove inorganic ash and shattered glass. The refuse is then passed through air classification (separation) to remove the dense fraction of the refuse stream. The light fraction is slurried in the

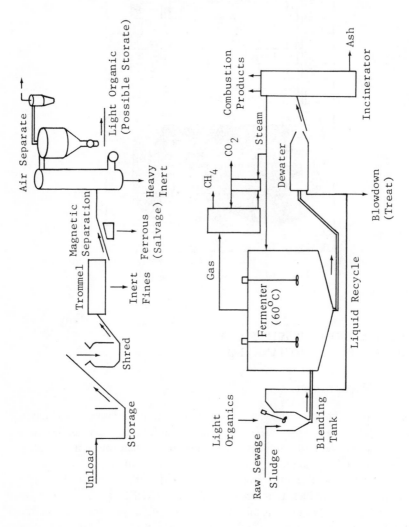

Figure 3-14.   Schematic of the process for producing methane from urban refuse [56].

blend tank to which recycled concentrate makeup water, raw sewage sludge, chemicals for pH control and nutrients, if required, are added. The slurry is conveyed to the fermentation tanks. The fermentation tanks are mixed anaerobic reactors operated at thermophilic temperatures of 55°-60°C.

The methane is separated from the carbon dioxide with the use of monoethanolamine as the reagent. The reactor slurry is dewatered by centrifugation, and the remaining concentrate is recycled while the cake containing 35-40% solids is incinerated.

The total methane gas production is a function of reactor temperature and volatile solids destruction. It is also dependent on the feedstock, hence a high-quality feedstock is desired to maximize gas production. The gas yield is substantially greater in the thermophilic temperature range, with a maximum of 0.45 m³/kg over 30 days of retention time. The rate of gas production from a system operating at a 4-day liquid retention time at 60°C is about 85% of that obtained at a 30-day retention time.

The residue incineration process is viewed as an asset since this reduces the cost of hauling residue to a landfill site. Secondly, the installation of heat recovery in the incineration system should provide more than enough steam to satisfy the process heating requirements. This improves the energy recovery efficiency. Table 3-5 shows energy balances in terms of caloric values for refuse from St. Louis, Missouri. It is evident that the residue has a higher energy value than the raw refuse, both in terms of total dry solids and volatile solids. This is attributable to the removal of cellulose from raw refuse while lignin with higher energy value is still present in the residue. The heat recovery is a function of excess air, furnace temperature and stack gas cooling.

This process has the potential to greatly reduce the cost of refuse disposal. The advantage of this process lies in the fact that a significant increase in energy efficiency can be obtained by incinerating the organic residues. The steam obtained by this heat recovery process will enable a recovery of approximately 60% of the energy input [56].

Table 3-5. Calorific Values for St. Louis Refuse [56]

	Volatile Solids (%)	Total Solids (J/kg)	Volatile Solids (J/kg)
Raw Refuse	66.4	14×10^6	21.6×10^6
Residue	77.1	18.5×10^6	24×10^6

Stage of Development

There is a demonstration plant in Pompano Beach, Florida.

Implications for Energy Consumption

Energy efficiency can be achieved by incinerating organic residues instead of using them for landfills. The energy needed to haul the urban refuse to landfill sites is quite substantial. When compared to the energy need for incineration there is substantial energy conservation.

USE OF BIOPOLYMER (XANTHAN GUM) IN ENHANCED OIL RECOVERY

Description

Exopolysaccharides are polysaccharides found outside the microbial cell wall and membrane and are a common product of microbial cells. Polysaccharides produced from microorganisms have advantages over those produced synthetically. Some of the advantages are as follows:

1. Polysaccharides produced from microorganisms are unaffected by marine pollution, tides, weather, war, famine or drought.
2. Production can be controlled within precise limits and the scale of production can be geared to the market.
3. Location of the production facility also can be arranged to utilize convenient or cheap substrates.

The one disadvantage is the high cost of installation and startup of fermentation equipment, together with large solvent requirements and the associated need for a considerable amount of energy. It appears that this disadvantage can be overcome given sufficient application, money and innovation.

The greatest single potential market for commercially produced exopolysaccharides is that provided by the oil industry. These polymers used in tertiary oil recovery improve water-flooding and micellar-polymer operations. The role of polymer is to reduce the flow capacity of the solution in the rock system. Xanthan, a bacterial exopolysaccharide from the genus *Xanthomones compestris*, has been shown to produce higher viscocity and low sensitivity to saline than the synthetic polymers. The effectiveness of an exopolysaccharide depends on the salinity, pH, temperature, viscosity and characteristics of an oil field. Currently, problems encountered in using Xanthan include plugging of the rock near wells due to the presence of particulate material (i.e., bacterial debris) and changes in the polysaccharide concentration due to chemical adsorption onto the permeable rock structure. Detailed research is underway to overcome these difficulties.

Use of Xanthan as a drilling mud has been patented by Exxon [57]. Table 3-6 presents other nonfood patented applications for *Xanthomonas compestris* polysaccharide and its derivatives [57].

The other applications of Xanthan in the food industry include stabilizing

Table 3-6. Patented Applications for *Xanthomonas Compestris*
Polysaccharide and Derivatives [57]

Application	Patent Holder	Patent Number
Oil-Drilling Muds	Exxon	U.S. 3251768 (1966)
Stabilizer–Emulsion Paints	Tenneco Chemicals	U.S. 3438915 (1969)
Stabilizer–Water-Based Paints	Kelco	U.S. 3481889 (1969)
	Heyden Newport Chemicals	French 1395294 (1965)
Suspending Agent for Laundry Starch	Henkel	U.S. 3692552 (1972)
Carrier for Agrochemicals	Kelco	U.S. 3717452 (1973)
Agricultural and Herbicidal Sprays	Kelco	Canada 806643 (1969) U.S. 3659026 (1972)
Metal Pickling Baths	Diamond Shamrock	U.S. 3594151 (1971)
Gelled Detergents	Chemed	U.S. 3655579 (1972)
Gelled Explosives	Kelco	U.S. 3326733 (1967)
Waterproof Dynamite	Ashland Oil and Refining	U.S. 3383307 (1968)
Clay Coatings for Paper Finishing	Kelco	U.S. 3279934 (1966)
Flocculant for Water Clarification	Ashland Oil and Refining	U.S. 3342732 (1967)
Derivatives		
Alkylene Glycol Ester for Foods, Cosmetics	Kelco	U.S. 3256271 (1966)
Carboxyalkyl Ethers for Paper, Textiles	Kelco	U.S. 3236831 (1966)
Dialkylaminoalkoxy Ethers for Sizes	Kelco	U.S. 3244695 (1966)
Hydroxyalkyl Ethers for Cosmetics	Kelco	U.S. 3349077 (1967)
Sulfate for Glue Thickening	Kelco	U.S. 3446796 (1969)

agents in french dressing, fruit-flavored beverages, processed cheese and other dairy products. Fermenters used in the batch fermentation have a 900-liter capacity and contain a culture medium of corn syrup, distiller's solubles and salts. The product recovery is facilitated by organic solvent precipitation, followed by certifugation to remove some of the bacterial cells. The production of Xanthan occurs principally during the first 72 hours of culture, while prolonged incubation results in further polymer production and consequent higher culture viscosity.

Improvements to the batch culture development of Xanthan include continuous culture [58]. The conventional method of microorganism-derived exopolysaccharide production uses batch cultures with relatively long incubation periods to ensure maximum production of polysaccharide and maximum utilization of substrate. Continuous culture technique development, in progress in Britain, is an option where output can be increased for a given size of fermenter production. In continuous culture, *X. compestris* has been shown to grow well in a simple salt medium with glucose as the sole carbon source. The continuous cultures could be run for up to 500 hours without the development of mutant strains.

Stage of Development

Pilot R&D is underway for continuous culture technique. Also, there is batch culture commercial production.

Implications for Energy Consumption

None are known at present.

BIOCONVERSION OF CELLULOSIC FIBER TO ETHANOL

Description

Plant biomass, available as the most abundant renewable resource, has been used extensively as a substrate to produce ethanol as a liquid fuel and chemical feedstock. The process involves enzymatic saccharification of the cellulose to glucose and subsequent fermentation by yeast as shown below:

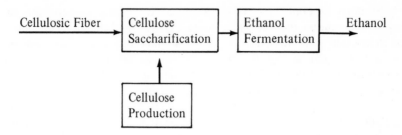

An alternative to this conventional process, described below, omits the steps of enzyme preparation and saccharification, thus producing ethanol by direct fermentation of plant biomass. The soluble sugar from the hydrolysis of cellulose is fermented in situ, thus minimizing inhibition of cellulase activity. This direct fermentation avoids the separate enzymes preparation and saccharification step shown above. In principle, direct fermentation of cellulosic fiber to ethanol can be achieved either by a mixed culture employing a cellulolytic anaerobe and a compatible ethanol producer, or by a pure culture of a cellulolytic ethanol producer (Figure 3-15).

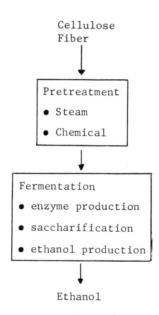

Figure 3-15.   Direct fermentation of plant biomass to produce ethanol [59].

Laboratory experiments have employed a cellulolytic, thermophilic an-aerobe, *Clostridium thermocellum*, strain LQ8. In the bench-top fermenter, the fermentation is started with 10%, 24-hour-old inoculum growing on the same medium as that in the fermenter. The pH is maintained at 7.4 by continuous addition of 3 M NaOH through a peristaltic pump controlled by an automatic pH controller. The degradation of cellulose proceeds after an initial lag time of about 20 hours. The biological culture has the capability to produce ethanol and acetic acid in a molar ratio of about 2 to 1. When added, additional cellulose is degraded without a lag period. The rapid decline in cellulose levels probably is due to the presence of an extracellular cellulose in the fermenter broth. The amount of ethanol produced is low due to the simultaneous production of acetic acid. Further research is needed to increase ethanol production levels by isolating new strains of microorganisms that produce ethanol as a sole end product.

Currently, the cost of producing ethanol by this process is high due to the lower volumetric production efficiency and ethanol concentration. These higher costs can be offset by the use of low-cost raw materials such as agricultural and municipal wastes. However, further work is needed in the area of culture development (isolation of new strains) and lignocellulosic fiber fermentation technology.

Stage of Development

Laboratory-scale research and development is underway.

Implications for Energy Consumption

These are not clear at present.

GASOHOL PRODUCTION BY COMBINED SACCHARIFICATION AND FERMENTATION

Description

The production of liquid fuels from renewable resources such as starch and sugars from grains has been studied extensively. Ethanol thus produced is already being used successfully and extensively in Brazil as a gasoline extender in the form of gasohol. The process described here makes use of wood chips from poplar hybrids in a hot, aqueous butanol treatment, followed by a combined saccharification and fermentation step using thermally stable cellulase and a thermophilic fermentation organism.

The process design, shown in Figure 3-16, is developed to reduce the high energy and economic costs of ethanol recovery. This is achieved by running the saccharification-fermentation step at high temperature and reduced pressure. This allows an ethanol condensate of 20% or greater and significantly reduces the total distillation costs for production of 95% ethanol. This combined saccharification-fermentation step uses thermally stable cellulase derived from *Thermoactinomyces* sp. and a thermophilic fermentation anaerobe, *Clostridium thermocellum*. The *Thermoactinomyces* sp. grows effectively at 55°C and concomitantly produces cellulase. The fermentation starts at a pH of 7.4, which drops to 7.2 during the first 6 hours and then to 6.7 over the next 6 hours, reaching a constant level at around 24 hours. This decrease in pH is suggested to be linked with the release of extracellular protein. The decrease in pH also describes an initial lag phase, an exponential phase and a stationary phase. The cellulolytic enzyme system of the *Thermoactinomyces* sp. is stable over a wide pH range and elevated temperatures [60]. The advantages of this enzyme system include (1) optimum cellulolytic activity produced within 24 hours of growth, (2) enzyme stability through the saccharification step, hence it can be recovered, and (3) enzyme system compartmentalization into extracellular CM-cellulase and avicelase activities and cell-associated β-glucosidase activity, which permits optimal use in saccharification.

Clostridium thermocellum, an alcohol-tolerant mutant, is a principal catalyst used during the combined saccharification-fermentation step, as shown in Figure 3-16.

Stage of Development

Laboratory-scale research and development is underway.

Implications for Energy Consumption

These are not clear at present.

FOOD-PROCESSING WASTE FERMENTATION
FOR METHANE PRODUCTION

Description

Anaerobic digestion of wastes resulting from agricultural production, food production and food processing is a preferential disposal means as compared to disposal by aerobic treatment or land application. The anaerobic digestion process has been used in municipal sewage treatment

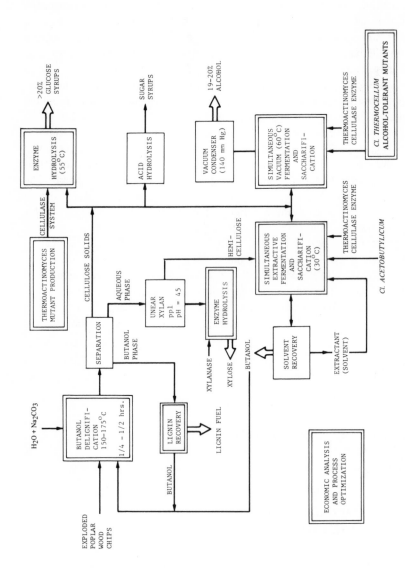

Figure 3-16.   University of Pennsylvania fuels from biomass process [60].

for many years, but its use in agriculture and industry has been limited. The reasons for limited use include:

1. relatively slow rate of conversion of organic substances to methane;
2. need for large digestors;
3. long retention time requirement; and
4. need to maintain temperature close to 35°C.

Recent modifications in digestion technology have helped to overcome these limitations. The laboratory process described here has enabled researchers to increase the rate of methane production at 35°C using the anaerobic process.

In the following laboratory-scale fermentation operation, each fermenter unit shown in Figure 3-17 is composed of a 30-liter cylindrical fermenter, a 6-liter settling flask, a refrigerated food tank, suitable peristaltic pumps and variable-speed stirrers. Food processing waste is pumped into the fermenter continuously at the desired rate. The effluents are overflowed from the settling flask at the same rate. Some of the food processing wastes obtained from commercial processing plants in concentrated forms are pretreated. Initially, the fermenter is inoculated with liquid obtained from an active sewage sludge digester.

The rate of methane production is dependent on the type of waste used and the nutrients added. Addition of complex bacterial nutrients in the form of yeast extract is known to increase/maximize the rate of methane production by a factor of 2–3.5, depending on type of waste. The COD:N:P ratio in the feed affects the rate of methane production, and about 300:5:1 is shown to be generally adequate. The feedrate is the rate-limiting factor in methane production.

The anaerobic contact process at 35°C is capable of producing over 2.0 m^3 methane/m^3/day from food processing plant wastes, with conversion efficiencies of up to 80%. Such high-rate methane production by anaerobic digestion makes this process competitive with methane obtained from dwindling fossil fuel sources. This process needs further research and refinement to answer the questions of stability, nutrition requirements, solids separation and the cellulose breakdown rate.

Stage of Development

Laboratory-scale research and development is underway.

Implications for Energy Consumption

These are not clear at present.

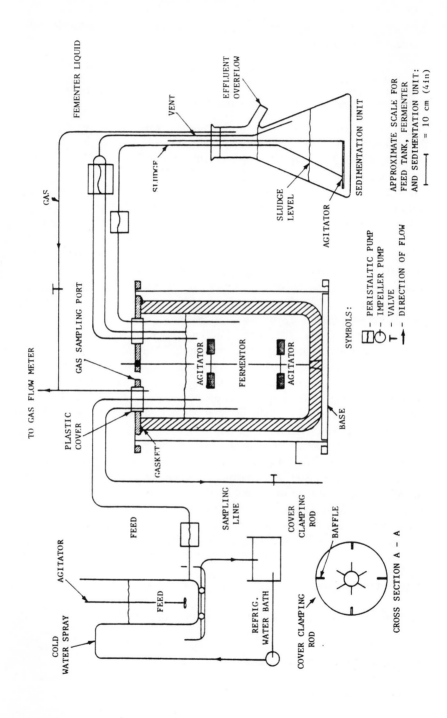

Notes: 1. Complete setup, except for refrigerated water bath, was kept in a temperature-controlled room (35°C, 95°F).

2. Fermenter consisted of 12-in. × 24-in. o.d. glass jar, with ground glass edge and 0.5-in. plastic cover. Baffles were 1.25 in. wide. Agitators were 4.75-in. diameter, with each blade measuring 0.5 in. × 1.75 in., and were mounted on a 0.5-in.-diameter rod turning in Teflon®* bearings both in the cover and at the bottom of the fermenter. Stirrer speed was variable from 50–800 rpm.

3. Sedimentation unit consisted of a 6-liter Erlenmeyer flask. Fermenter liquid inlet tube was vented to avoid disturbing effluent layer by gas bubbles. Agitation was provided by a single-bladed rubber impeller (4 in. long, tapering from 0.5 in. wide at shaft to about 0.25 in. wide at the end), rotating at 30–45 rpm, depending on suspended solids content.

4. The level in the fermenter was controlled within narrow limits by the position of the inlet tubes to the pump transferring fermenter liquid to the sedimentation unit. The level in the sedimentation unit was controlled by the overflow outlet.

*Registered trademark of E.I. duPont de Nemours & Company, Inc., Wilmington, Delaware.

Figure 3-17.   Schematic outline of fermenter setup [61, 62].

METHANE PRODUCTION FROM SEWAGE SLUDGE

Description

Conversion of organic waste to methane by anaerobic digestion has several advantages over the other gasification processes. Most types of high-moisture-content organic feed can be used for the production of intermediate-Btu gases under low temperature or pressure conditions at relatively high overall thermal efficiencies. Another advantage is that ungasified solid and liquid end products of anaerobic digestion can be utilized as sources of nutrients without undue environmental impacts.

The process described here is a two-phase anaerobic digestion designed to optimize process configuration based on the biochemical reaction kinetics and consists of separate acid and methane digestors. The laboratory-scale system is operated at 37°C with 90% activated sludge and 10% primary sludge mixtures. Two 10-liter plexiglass digestors are operated in tandem without organism recycling. The acid is fed continuously with sludge in the digestor, and the acidic effluent from this digester is fed on a daily basis to a methane digestor. Figure 3-18 shows the two-phase digestion of cellulosic feed.

This laboratory-scale, two-phase system has demonstrated the following:

1. It is possible to separate enrichment cultures of acidogenic and methano-genic organisms by kinetic control involving manipulation of dilution rates and the microbial generation time.
2. Hydrolysis and acidification of sewage sludge are the predominant reactions in the acid-phase digestor.
3. The two-phase system showed a volatile solid reduction efficiency of 40%, gas and methane yields of 15.7 and 10.7 scf/lb of volatile solids, and a retention time of 6.46 days. This type of performance is to be compared with that in a conventional digestor with retention times of 14 days or longer.

Volatile solids reduction, methane production and thermal efficiency can be expected to increase compared with conventional digestors [63]. Clearly, the two-phase mode of operation can result in savings in capital and operating costs of biological methane production from organic feeds.

Stage of Development

Laboratory-scale research and development are underway.

Implications for Energy Consumption

The energy requirement is reduced due to shortening of the detention time; further, there are reduced heating and mixing requirements.

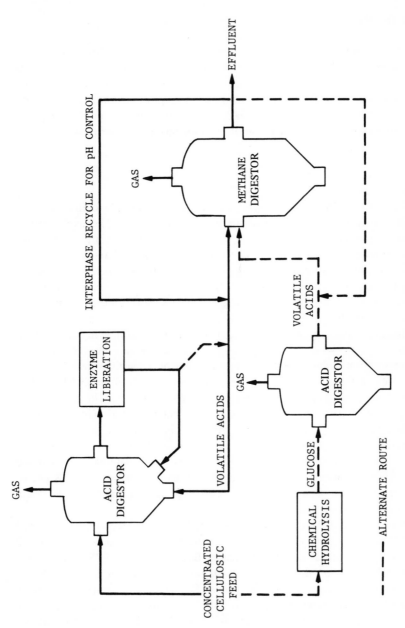

Figure 3-18. Two-phase digestion of cellulosic feed [63].

HYDROGEN GAS PRODUCTION UTILIZING ORGANIC SUBSTRATE

Description

One potential method of producing energy in the form of hydrogen gas is to use microorganisms capable of oxidizing an organic compound in the presence of light. The bacterium known to do this is *Rhodospirillium rubrum*, a nonsulfur purple bacterium. The process outlined below uses immobilized *R. rubrum* and malate as a substrate to produce H_2 in a continuous process. The reaction of this bioconversion is as follows:

$$C_4H_6O_5(\text{Malate}) + 3H_2O \rightarrow 4CO_2 + 6H_2$$

In the laboratory, *R. rubrum* is grown in a chemically defined medium and stored in the spent medium at 4°C until used. The experiment is carried out in a box filled with argon gas. After preparation, the flasks are stoppered and incubated 60–95 minutes on a mechanical shaker with constant illumination provided by two rows of 100-W standard incandescent light bulbs. Aliquots of gas (usually 0.2–1.0 ml) are removed from the flasks and assayed for hydrogen, nitrogen and oxygen production on a gas chromatograph.

Immobilization of *R. rubrum* is achieved by the following steps. A 4-g amount (wet weight) is added to 15 ml of Noble Agar 5% solution. This solution is spread evenly on both sides of a plastic slab composed of a series of thin plastic layers. The slab is then put into an airtight system. The system is immersed in water maintained at 18-19°C. The gas produced is channeled to a double-ended glass tube that is kept airtight by septums. The malate solution is pulled through the collector with a syringe, displacing any gas with substrate. Collection of the produced gas displaces the liquid. A sample is taken after the gas and the malate have equilibrated, and then a constant flowrate is reestablished. Illumination for this process is provided by a rack of seven 100-Watt standard incandescent bulbs on each side of the reactor.

Based on this laboratory experiment, researchers feel that whole cells of *R. rubrum* can be immobilized in an active form in which they will produce H_2 on a continuous basis. Based on the 1976 dollar value, researchers suggest that this process is not economical at a cost of $10/lb of malate. This works out to a cost $8.33/1000 ft^3 [64]. The heating capacity of H_2 is about one third that of natural gas vol/vol. Thus the price is equivalent to $25/1000 ft^3 natural gas based on raw materials (substrate) cost only.

Stage of Development

Laboratory-scale research and development is underway.

Implications for Energy Consumption

Based on 1976 figures, it is uneconomical due to high raw material costs and resultant H_2 obtained per pound of substrate.

HYDROGEN PRODUCTION THROUGH BIOPHOTOLYSIS

Description

Biophotolysis is one of the approaches through which hydrogen, a nonpolluting fuel, could be produced by means of solar radiation. Table 3-7 presents proposed biophotolysis systems. The most advanced system developed to date is based on the use of nitrogen-limited cultures of heterocystous blue-green algae. Figure 3-19 shows a model of metabolism by heterocystous blue-green algae.

In this system, H_2 and O_2 evolving reactions are separated at the microscopic level in the heterocysts and vegetative cells of these filamentous

Table 3-7. Proposed Biophotolysis Systems [65]

		Status[a]
Single-Stage Systems		
1	Chloroplast−ferredoxin−hydrogenase	Laboratory
2	Primary electron acceptor	Conceptual
3	Heterocystous blue-green algae	Outdoor
4	Alternating H_2 and O_2 evolution	Conceptual
Two-Stage Systems		
5	Oxygen-stable intermediate	Laboratory
6	Alga culture cycles between stages	Conceptual
7	Organic wastes/photofermentations	Laboratory
8	Reversible oxygen trap	Experimental
9	Membrane separation of reactions	Conceptual

[a]Conceptual: no experimental work, based on known biochemistry and physiology.
Experimental: not yet demonstrated; however, laboratory experiments are in progress.
Laboratory: shown in the laboratory; not yet scaled up.
Outdoor: demonstrated with a model converter under outdoor conditions.

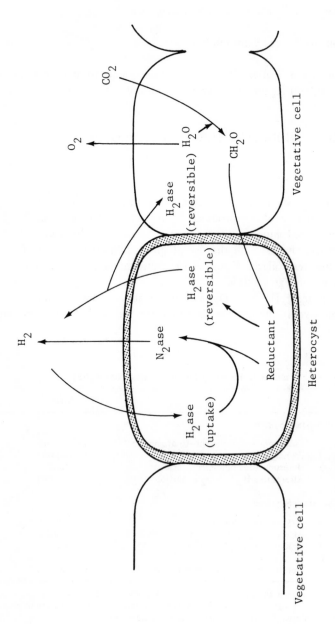

Figure 3-19. Model of hydrogen metabolism by heterocystous blue-green algae [65].

blue-green algae. Nitrogen-limited cultures of *Anabaena cylindrica* have been shown to evolve H_2 up to four weeks, both in the laboratory and outdoors.

In the laboratory equipment, *Anabaena cylindrica* 629 is grown aerobically in a 45-liter tank illuminated on both sides by six 40-watt fluorescent bulbs. The culture temperature is maintained within the range of 26°C to 29°C and the pH controlled at 7.5±0.5 by adjusting the carbon dioxide (CO_2) content of the cell by sparging the air. The concentrated cultures are transferred outdoors when the cell density of 0.35 mg/ml is obtained. Heterocysts are induced by sparging with N_2/CO_2/argon (0.2–0.4%, 0.5%, balance). Hydrogen production can be observed within one to four days after the transfer, depending on both cell density and solar radiation.

The outdoor converter is an array of 1-liter glass columns (5 cm in diameter, with 0.8 liter of working volume) and is placed at a 35° angle to the horizontal facing south.

The biophotolysis reaction gives a discontinuous production of H_2 because of the diurnal cycle. It has been demonstrated that H_2 production activities respond very rapidly to the start of sunshine in the morning, reaching a maximum within 30 minutes of direct sunlight on the converter. A similar reverse pattern is demonstrated at sunset. The limiting factors in H_2 production are the length of the algal filaments in the cultures, N_2-level and presence of reductant in the system.

Photosynthetic bacteria are known to produce energy in the form of hydrogen from an oxidizable organic source, but they are incapable of oxidizing water. Hence, no O_2 is evolved in bacterial photosynthesis. On the other hand, blue-green algae do produce O_2. In this laboratory system, the photosynthetic bacterium *Rhodospirillum rubrum* and the blue-green algae *Anacystis nidulaus* are combined into a single system, with resultant production of H_2 plus O_2 by undergoing the following reaction:

$$A.\ nidulaus \quad 2H_2O + 2NADP^+ \rightarrow O_2 + 2H^+ + 2NADPH$$

$$R.\ rubrum \quad 2NADPH + 2H^+ \rightarrow 2NADP + 2H_2O$$

Anacystis nidulens can produce NADPH from exogenously supplied NADP by the oxidation of water in a light reaction. This reduced NADPH can be reoxidized by *Rhodospirillum rubrum* to produce H_2.

A laboratory system is devised with two reactors using immobilized microorganisms. The two reactors serve the purpose of separating the oxygen-producing system from the hydrogen-producing system. The exogenous NADP from the reservoir continuously purged with argon, is pumped through the algae reactor where NADP is reduced. As it is produced, the NADPH is pumped to the *R. rubrum* reactor where the NADPH is reoxidized and H_2 is evolved. The reoxidized NADP is then recycled. The efficiency of green plants to convert solar energy to hydrogen ranges from 1.0 to 1.2%. It is suggested that this system compares favorably with the other biophotolysis systems [66].

Stage of Development

Laboratory-scale research and development is underway.

Implications for Energy Consumption

No external energy is required.

HYDROGEN GAS FROM CHEESE WHEY WASTE

Description

A variety of algae (blue-green and purple) and bacteria produce hydrogen. Before this hydrogen can be used as a fuel source, it has to be separated from oxygen and other gases produced by these organisms. This is generally an energy-consuming process. The light-dependent production of hydrogen by photosynthetic bacteria offers a potential solution to this problem during this process, since CO_2 is the only additional gas evolved. Algae are known to use water as a hydrogen donor whereas purple nonsulfur bacteria require an organic substrate. In the work summarized below, it has been shown that a high yield of hydrogen can be achieved using lactic acid-containing cheese waste as both a hydrogen donor and carbon source for photosynthetic bacteria.

In this laboratory system, *Rhodospirillum rubrum* is the light-dependent photosynthetic bacterium that utilizes lactic acid waste substrates. The cultures are incubated anaerobically in a closed vessel at 30°C under continuous illumination. A fermenter with a rectangular cross section and a volume of 1000 ml is used. Figure 3-20 illustrates the fermenter setup. The fermenter is illuminated by a 4-cm light path from the large side by two 100-Watt spotlight tungsten lamps. The culture is maintained at 30°C and mixed with a stirrer. The fermenter has a gas outlet from which a gas sample is withdrawn. Hydrogen production is continuous over a 40-day period. In this laboratory experiment, production of H_2 of 65 ml/hr/ℓ of culture volume is achieved. These results are comparable with the hydrogen production rates obtained with the blue-green alga *Anabaena*. It is suggested that by optimizing experimental conditions

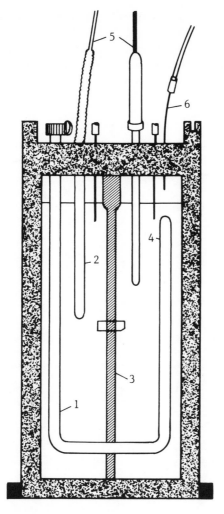

1. Cooling coil

2. Pt – 100 temperature sensor.

3. Stirrer.

4. pH electrode.

5. Inlet and outlet for growth medium and cells.

6. Gas outlet.

Figure 3-20.   Fermenter (1000 ml) with a rectangular cross section (4 × 12 cm) illuminated by a light path (4 cm) [67].

with *R. rubrum* it should be possible to reach rates of H_2 as high as 130 ml/hr/ℓ of culture [67].

A technique such as this offers the advantage of hydrogen production by utilizing a readily available industrial waste. Further work is needed, however, to produce hydrogen fuel in an economically feasible system.

Stage of Development

Laboratory-scale research and development is underway.

Implications for Energy Consumption

These are not clear at present.

CHAPTER 4

WASTE TREATMENT

This chapter discusses processes that are related to the treatment of waste materials.

USE OF DENITRIFYING BACTERIA IN MUNICIPAL WASTEWATER TREATMENT

Description

Nitrogen removal from wastewater is of concern since nitrogen is (1) a major cause of the eutrophication process, and (2) a potential danger to public health if present in excess in drinking water [68]. The conventional means of nitrogen removal include aeration, sedimentation, breakdown chlorination, selective ion exchange and nitrification-denitrification, in which a nonfluidized biological reactor is used. The pilot-scale process described below is capable of denitrifying wastewater before recharge.

This process of denitrification is a biological process in which bacteria reduce nitrite or nitrate in the influent waste stream to nitrogen gas, a non-polluting end product. Wastewater is passed upward in a cylindrical reactor through a bed of small particles, such as activated carbon or sand, at a velocity sufficient to cause motion or fluidization of the medium. The medium serves as a support surface on which the bacteria grow. As the wastewater is passed through the reactor, the bacteria in the medium reduce nitrogen-containing compounds to nitrogen gas. Bacteria are capable of growing on activated carbon or sand, thereby facilitating removal of contaminants in a given volume of reactor. This technique of fluidization increases the effective surface area when compared with the conventional parallel bed, thus minimizing operational costs and problems. Another advantage is that fluidization also allows a greater volume of waste to be treated in the reactor per unit time.

The bacteria cultivated in the reactor are heterotrophic, requiring organic sources of carbon for growth. This carbon source is supplemented in the reactor system by methanol. Under aerobic conditions, the greatest level of nitrogen reduction is obtained since the available nitrate is not used as an energy source. In other words, oxygen necessary for the respiratory process is available; hence, the nitrates are not reduced to provide oxygen. The nitrogen reduction is a respiratory process encompassing the following chemical reactions:

$$NO_3 + \frac{1}{3} CH_3OH \rightarrow NO_2 + CO_2 + \frac{2}{3} H_2O$$

$$\frac{NO_2 + \frac{1}{2} CH_3OH \rightarrow \frac{1}{2} N_2 + \frac{1}{2} CO_2 + \frac{1}{2} H_2O + OH}{Net: NO_3 + \frac{5}{6} CH_3OH \rightarrow \frac{1}{2} N_2 + \frac{5}{6} CO_2 + \frac{7}{6} H_2O + OH}$$

The denitrification reactor illustrated in Figure 4.1 consists of five 0.25-in. (0.6 cm)-thick plexiglass sections 1.5 ft (0.46 cm) in diameter, giving the column a total height of 15.5 ft (4.7 meters) and an empty reactor volume of 180 gallons (680 liters). A 0.125-in. (0.3 cm)-thick steel distribution plate is located 1.5 ft (0.46 m) up from the base of the column; on top of this are 1.5 in. (3.8 cm) of pea gravel and about 3 ft (1 meter) of white silica sand with an effective size of 0.6 mm, which serves as a nucleus for bacterial growth.

When the sand is covered with growth, the height of the fluidized bed is about 12 ft (3.7 meters). The rotary mixer at the top of the column is used for controlling the height of the bacteria-coated sand bed. The waste water enters at the bottom of the column and is passed upward through the fluidized bed just below the effluent outlet located about 1 ft (0.3 meter) from the top of the column. Figure 4-2 is a schematic of the nitrification-denitrification processes. The nitrate concentration in the influent is maintained by addition of sodium nitrate. The rotary mixer at the top of the column can be adjusted to control the bacterial growth in the fluidized bed. Any shutdown of the reactor also slows the bacterial growth.

The results of this pilot-plant operation have demonstrated that a fluidized-bed reactor can consistently produce an effluent free of nitrate and nitrite. Compared to the conventional packed-bed columns, this fluidized-bed reactor has been demonstrated as a relatively trouble-free operation without such problems as clogging or backwash [68]. This reactor works best under

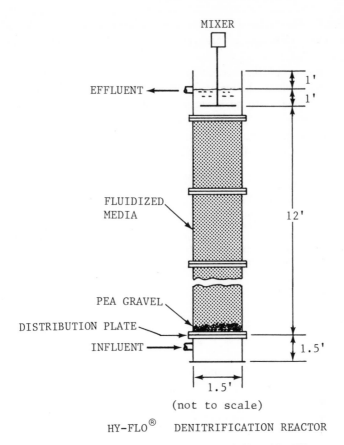

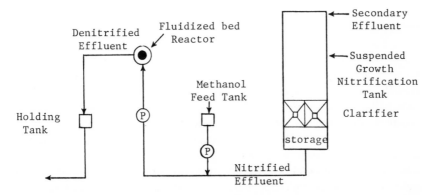

Figure 4-1. Denitrification reactor (ft × 0.305 = M) [68].

Figure 4-2. Schematic of nitrification-denitrification pilot plant [68].

constant flow conditions. The reactor has been shown to work optimally with at least 95% nitrogen removal capacity when the methanol:nitrate-N ratio is 2:9.

The efficiency of this system can be attributed to the concentration of active biomass within the reactor. The average bacterial concentration in the reactors is estimated to be between 30,000 and 40,000 mg/l. This concentration, which is 10 to 20 times greater than that in conventional biological systems, allows the reduction in time required for denitrification. The advantages of the fluidized-bed reactors are as follows [68]:

- capacity of 36,000 gpd (136,000 liter/day) of waste water,
- rapid nitrogen removal (99%) in less than 6.5 min at a flux rate of 15 gpm/ft^2 (620 liter/min/m^2),
- simple operational routine,
- removal of concentrations of nitrate much greater than would be expected in a typical municipal plant, and
- potential energy savings.

Stage of Development

There is a pilot plant at Bay Park Water Renovation Plant, Nassau County, New York.

Implications for Energy Consumption

Nitrogen removal is achieved in less time than conventional methods, thereby reducing energy consumption.

USE OF MUTANT BACTERIA IN TREATMENT OF OILY WASTES

Description

The addition of mutant bacteria can speed the restart and enhance the overall efficiency of oil refinery biological waste treatment systems. HYDROBAC® is a formulation of mutant, adapted microbes and biochemical accelerators formulated specifically for petroleum refinery/petrochemical plant waste waters. This product degrades various hydrocarbons and organic chemicals that may be toxic, inhibitory or bioresistant to natural microbial populations, thus stabilizing the operations of biological waste water and waste disposal systems that are subject both to changes in incoming waste concentration and composition and to variations in waste treating conditions such as pH, temperature and nutrient levels.

Refinery/petrochemical wastewaters typically contain chemicals ranging from readily degradable to highly toxic substances. This wide spectrum of substrates and the varied operating conditions (pH, temperature and flow) create stress on the natural biomass of biological wastewater treatment systems and waste disposal systems and result in lowered efficiency of natural biomass for degrading the applied waste. In activated sludge plants these variations are known to hamper the operations of secondary clarification equipment. This phenomenon, called biomass stress, also tends to cause foaming within the wastewater treatment system.

Figure 4-3 demonstrates a refinery wastewater treatment system using HYDROBAC degrader. The activated treatment sludge unit has two parallel treatment trains, each consisting of an aeration basin, a clarifier and a sludge recycle system.

HYDROBAC is a mutant bacteria, which, working in conjunction with the indigenous biomass, can rapidly degrade toxic, inhibitory and otherwise bioresistant organic wastes. The microorganisms in this commercially available bacterial mixture can biologically degrade benzenes, phenols, cresols, napthhalenes, amines, alcohols, synthetic detergents, petroleum (crude and processed), kerosene, cyanides and other bioresistant wastes from refineries [69].

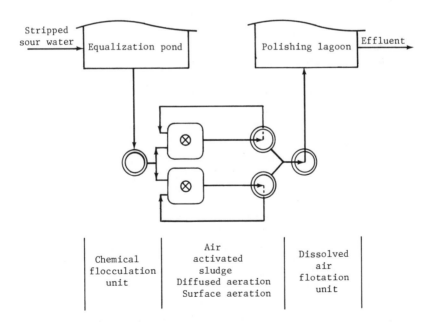

Figure 4-3.   Refinery wastewater treatment system [69].

HYDROBAC also can be used to augment the performance of oily sludge farming operations and hazardous materials spill cleanup operations, as well as to improve and expand treatment plant operations. HYDROBAC has been used successfully to clean up spills of such bioresistent compounds as ortho-chlorophenol, dioxin, diesel fuel and acrylonitrile [69].

Stage of Development

Mutant bacterial cultures are available commercially from Polybac Corp., 1251 S. Cedar Crest Blvd., Allentown, Pa.

Implications for Energy Consumption

There is a general increase in treatment plant efficiency.

IMPROVED OXYGEN UTILIZATION IN WASTEWATER TREATMENT BY THE AERATION TOWER TECHNOLOGY

Description

In conventional horizontal treatment basins diffused-air aerators are used. These are rectangular concrete tanks fitted with perforated devices near the bottom. The compressed air is injected to produce air bubbles, which, on rising through the water, produce turbulence. For deeper tanks, higher air compression is required, which adds to the operation costs since the rising bubbles expand and tend to coalesce. The disadvantages of conventional horizontal treatment basin technology center on space and energy require-ments, as well as poor utilization of oxygen. The technology known as Bayer's tower biology is designed to overcome these disadvantages.

Bayer's tower biology system offers a saving in space because of its vertical structure. In addition, it enables improved oxygen utilization and reduced offgas treatment, resulting in energy savings of better than 50% and capital cost reductions of up to 20% [70]. The Bayer system has a potential use as an add-on treatment step within the existing industrial facility.

In Bayer's tower biology system, shown in Figure 4-4, wastewater and air are injected into the bottom of a steel tower through specially designed nozzles. The kinetic energy of the water stream breaks the air stream into small bubbles, thus maximizing surface area. This maximization of surface area results in increasing the reaction potential between the bacteria and wastewater bubbles. This, in turn, results in improved oxygen utilization by

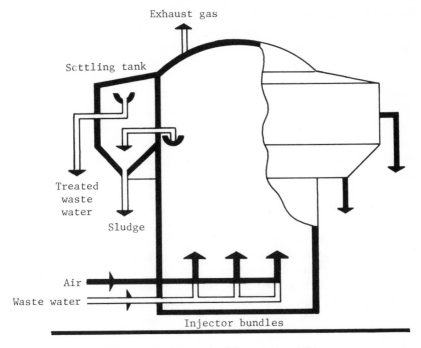

Figure 4-4.   Schematic of Bayer system [70].

the aerobic bacteria and secondary treatment quality, with over 90% reduction in BOD and suspended solids. As the bubbles rise in the tower, they sustain the bacteria consuming the organic material in the waste. The treated wastes overflow near the top of the tower and flow into a settling tank, suspended as a ring around the top of the tower. Sludge is drawn off the bottom and treated water off the top.

The height of the column is the critical factor in this system. It has been demonstrated that beyond a height of 100 ft (30.5 meters) the cost of pumping is prohibitive, even though there is increased oxygen utilization. With this tower system only 8% of the oxygen remains unused, compared to 16% in injector-aerated horizontal basins. Improved oxygen utilization means less air is used, which effectively lowers the amount of exhaust gas from the top of the tank. This results in lower fuel requirements for treating the gas and incineration of sludge. These two factors contribute to the greater than 50% energy savings over horizontal basins, even though a considerable amount of energy is needed to pump the wastewater into the system.

Stage of Development

A pilot plant has been operating for 1.5 years. A full-scale plant was to be completed by 1980 in Leverkrisen, Germany. The full-scale unit has four towers, each 85 ft (25.9 meters) high and 85 ft (25.9 meters) in diameter.

Other companies using the same concept for biological treatment of waste water include the following: (1) deep shaft system, Imperial Chemical Industries, England; (2) Carrousel, BASF, West Germany; (3) Electrolux A.B., Sweden; and (4) Ecolotrol, Inc.'s fluidized-bed version manufactured and marketed by Dorr-Oliver, Stanford, Connecticut.

Implications for Energy Consumption

Power consumption is reduced by 50%, compared to conventional pond aeration due to improved oxygen utilization. Due to the greater pressure in the tank and specialized inspector system, finer bubbles are produced in the tower. This results in a larger overall bubble surface area. Because small bubbles rise more slowly than large ones, waste bubbles reside in the tower longer, thus facilitating longer exposure of the aerobic bacteria to the organic waste.

MUNICIPAL SEWER SYSTEMS CLEANUP USING GREASE-EATING BACTERIA

Description

The flow in a sewer system is reduced when mineral and other particles combine with grease on the inner surfaces of the sewer pipes. This is believed to be a major contributor to sewer backups, overflow, foul odors and related pollution problems.

Traditionally, cleaning of sewers is accomplished mechanically using rodding and bucket machines, high-velocity waterjets and hydraulically propelled tools and devices. These techniques are expensive in terms of materials and energy costs. The new method described below utilizes grease-consuming bacteria for sewer cleaning. This method was developed for use in the sewage system served by the Washington Suburban Sanitary Commission, which is 3000 miles long and has an operating capacity of 150 million gpd. It is estimated that cleaning such sewers by traditional means requires approximately $4 million annually [71].

Grease-consuming bacteria, similar to the "oil-eating" bacteria used in the cleanup of oil spills, are cultured on an ample supply of food (i.e., grease). Following this growth cycle, the food supply is discontinued and the bacterial mass is allowed to starve (fast), thus producing a culture of healthy bacteria. After several feed and starvation cycles, the bacterial culture is freeze-dried until further use.

Before application, the bacterial culture is mixed with water, and activated bacterial solution is added to sewer mains and laterals via manhole covers. These bacteria multiply by consuming greasy material on the sewer pipes. When the bulk of the accumulated grease in the sewer system is removed, the bacterial population decreases because of the reduced supply of food. A continuous low dose of bacterial culture is then added to maintain the grease-free condition in the sewer system.

A sewage system cleanup performed in this manner has the potential for improved efficiency and safety over conventional methods. It is estimated that the annual cost of cleaning by bacterial culture is approximately $1.3 million (to be contrasted with the estimated $4 million annual cost for conventional methods) [71].

Table 4-1 summarizes the experience of several users of bacterial cultures for sewage system improvement. The removal of grease from lift stations and continuous force mains reduces the energy required to operate the station pumps and, thus, the total energy consumption of the sewer system. From Table 4-1, it is evident that the use of bacterial culture in the Erie, Pennsylvania, sewage system has reduced the energy consumption of the station by about a factor of three. In Arvada, Colorado, bacterial cleanup of the digestor at the sewage treatment plant resulted in approximately a threefold increase in methane gas production. This offers an incentive to use bacteria for grease cleanup in sewage treatment systems as well as in sewer lines.

Stage of Development

The technique is currently being used in various facilities in the United States. Necessary products (bacterial cultures) are commercially available.

Implications for Energy Consumption

There is substantial energy reduction (by a factor of three) compared to the conventional methods, and annual operating costs are reduced by 60%.

USE OF OXIDATION PONDS IN WASTEWATER TREATMENT

Description

Oxidation ponds are currently the subject of research as potential tools for both primary (solid removal) and secondary (oxidation) treatment of

Table 4-1. Survey of

Name and Address of Company	Population Served	Area	Total Length of Sewerage System
Ralph Shook Yorba Linda Water Dist. Yorba Linda, CA 92686 (714) 528-6231	2,500	N/A	N/A
Clearance Cardener Water & Sewer Superintendant 80001 Ralston Road Arvada, CO 80002 (303) 426-6441	70,000	Arvada, CO	280 miles
Wayne Coloney, P. E. P. O. Drawer 3966 Tallahassee, FL 32303 (904) 385-8171	10,000	City of Perry, FL	20 – 30 miles
Morris C. Allen Tahoe City Public Utility Tahoe City, CA 95730 (916) 583-3796	25,000 Average 75,000 Peak	Tahoe City, CA	City
Russ Bond, District Supt. Laguna City Sanitation P. O. Box 1068 Santa Maria, CA 93454 (805) 925-1475	19,000	Laguna County, CA	N/A
Paul Cygan, Chief Bureau of Sewers Dept. of Public Works City of Erie Erie, PA 16512 (814) 456-8561	180,000	Erie and outlying community	N/A
Charles Caswell, President Environmental Audit Corp. 5100 Center Avenue Pittsburgh, PA 15232 (412) 682-1031	N/A	Chicago	N/A
W. E. Cochrane Trussville Sewage Treatment Plant Rt. 2, Box 1-A Trussville, AL (205) 655-3617	3,000	Trussville	N/A

Bacteria Users [71]

Type of System	Capacity	Comments
Collect and transport to regional system	N/A	Bacteria cultures were used successfully to remove localized grease buildup.
Secondary with chlorination; trickling filter with two-stage digestor	1 mgd	Symptoms indicating inefficient digestor operation were: (a) inadequate production of methane; (b) 4-6 ft blanket in the digestor; (c) sludge had excessive amounts of organic and volatile contents. Within 30-40 days gas production doubled, tripled in 60. Also, there was a tremendous improvement in the sludge.
Primary	0.75 mgd, dry day; 1.25 mgd, wet day (badly infiltrated system)	This system operated at 28–30% BOD_5 and suspended solids (SS) removal. In the first year of bacteria use, BOD and SS removal increased 50–70%, the second year to 75%. A polishing pond increased this to 95%. Conventional treatment would have cost \$300,000, while the pond cost \$22,000 and \$200/month for bacteria.
Primary	3.5 mgd	This plant uses 1.5 lb/day of bacteria culture, costing about 2.5¢/million gallons. Tremendous decrease in odor, grease, solids and BOD was observed.
Secondary, filter	1.25 mgd	To remove grease and scum, 1 lb/day bacteria culture was used in the digestor.
Activated sludge	65 mgd	Bacteria cultures have been used very successfully for several years in cleanup of lift stations, scum wells and lines, skimmer lines, primary and aeration tanks. Lift station energy requirements were reduced by factor of 3 and offensive odors eliminated.
	0.108 mgd	Bacteria cultures were used primarily for removal of grease and industrial waste from oil tankers. Treatment was 1 lb every other day, gradually reduced to 1 lb every week. Prior to use, grease content was 200-300 ppm; after application, oil content dropped to 0-3 ppm.
Bowl filter	0.4 mgd	The city's old filter was malfunctioning with excessive pounding. Bacteria cultures were added and significant improvement was noted within six weeks, completely clearing the filter and stopping the pounding. In 13 weeks the digestor began to produce methane gas.

municipal wastewater. In these ponds, microalgae grow and produce the oxygen required for bacterial breakdown of the biodegradable, organic wastes. These ponds are predicted to have lower capital, operating and energy costs than more conventional, widely used wastewater treatment processes such as the activated sludge process. One disadvantage is that the high concentration of microalgae in the effluents does not meet current U.S. Environmental Protection Agency (EPA) standards of a maximum of 30 mg/l.

The oxidation pond effluents produced after microalgae removal by microstraining or in-pond settling contain sufficient nutrients to allow the growth of nitrogen-fixing blue-green algae. These algae have the unique capability of simultaneously generating oxygen photosynthetically while reducing atmospheric nitrogen (N_2) to ammonia. Figure 4-5 shows the concept of such a microalgae system, which serves the multiple function of advanced wastewater treatment—water reclamation, net fuel production, and fertilizer recycling and production. In this multistage system, the first stage consists of growing green algae up to, or approaching, the limit of readily available nitrogen (ammonia). In the second stage, remaining ammonia is utilized and the algae are then settled out in a batch growth or isolation process. The effluent from this system is used to cultivate a crop of nitrogen-fixing blue-green algae, while the microalgae biomass is a suitable substrate for methane production by anaerobic digestion.

The oxidation ponds mixed by paddle wheels are either rectangular or circular and are operated at depths of 20–30 cm. The objective of the continuous-growth ponds is to produce an effluent suitable for cultivation of blue-green algae. This is accomplished by limiting nitrogen availability necessary for the growth of green algae. Low nitrogen levels are sufficient to allow blue-green algae growth.

Ammonia, a nitrogen source for the growth of green algae, is depleted by two processes: uptake by algae and outgassing. *Anabaena* and *Nostoc* are the predominant algae in the cultivated culture, while *Oscillatoria*, diatoms and some green algae are contaminants. Cultivation of blue-green algae is temperature dependent. Hence, the technique outlined here can only be used in warmer climates. Cold-adapted strains need to be developed that can be used in outdoor cultivation. The factors that could affect pond operation include increased sewage concentration, changes in algal composition and rate of ammonia removal.

The year-round reliability of these oxidative pond systems is not yet established. This system, as a method of tertiary waste treatment combined with fuel and fertilizer production, has obvious merits. The low energy requirement could potentially reduce energy costs, which are customarily high for conventional wastewater treatment systems. The energy requirement for this system is for operation of pumps, blowers and paddle wheels.

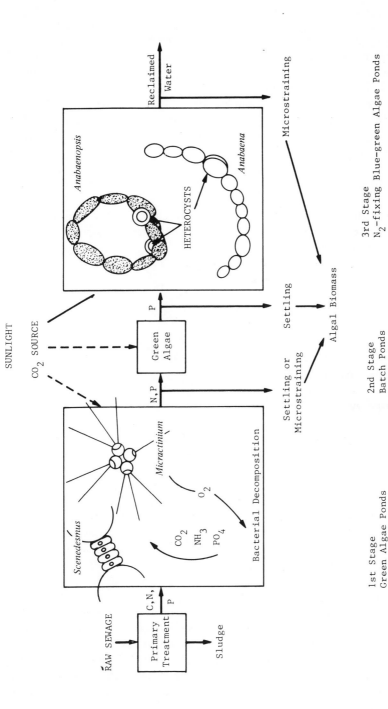

Figure 4-5. Advanced wastewater treatment using a multistage ponding system [72].

Removal of biomass or harvesting by microstraining offers reduction in energy requirements amounting to about 70 kWh/mg for 100-mgd plants. (1 mgd corresponds to a plant typically serving 10,000 people.) In relation to conventional wastewater treatment, such costs for algae harvesting make oxidation ponds competitive. The technology of microalgae oxidation ponds for sewage treatment can be adopted for near-term development of biomass for fuel production through anaerobic digestion to produce methane. The production of chemical feedstocks, liquid fuels (e.g., ethanol) and fertilizer is also possible.

Stage of Development

Laboratory-scale research and development is underway involving blue-green algae cultivation in oxidation ponds. Microstrainers currently are used in wastewater treatment.

Implications for Energy Consumption

There is a reduction in energy use, as described above.

REMOVAL OF SULFUR COMPOUNDS FROM INDUSTRIAL WASTEWATERS

Description

The presence of sulfur-containing materials in a variety of industrial effluents may be due to impurities in raw materials or process by-product formation. The total production of such waste poses a disposal problem since it may be toxic to the organisms in activated sludge. It has been suggested that if these materials can be reduced to less toxic forms by microbiological pretreatment, then municipal activated sludge treatment can be used effectively for further purification. Some trends that show potential for use in microbial pretreatment are outlined here.

Sulfur dioxide (SO_2), hydrogen sulfide (H_2S), methyl sulfide and ferrous sulfate are some of the sulfur-containing compounds found in the industrial waste materials of coal, oil and pulp industries. The environmental effects of these are toxic and corrosive in nature. Microbial pretreatment systems are being developed using *Thiobacilli*, which are tolerant to hydrogen sulfide and methyl sulfides. These microorganisms are capable of oxidizing volatile sulfides to sulfates and elemental sulfur. An industrial-scale operation, consisting of three 500-m^3 units, operates under variable conditions of temperature, pH, nutrients and aerobiosis (i.e., number of living *Thiobacilli*).

A fully efficient operation is preceded by a lag period that represents the establishment of a microbial population suited to the selective conditions. It has been demonstrated that in full operation, 90–95% of the volatile sulfides may be removed from effluents and waste gases by this method, with an effluent loading rate of 2 m^3 effluent/m^3/day.

Such a microbial pretreatment of industrial effluents by highly specific, efficient microorganisms resistant to toxic chemicals potentially can render less toxic forms. These less hazardous and less toxic effluents may be more suitable for the conventional purification processes involving activated sludge. These sulfur-oxidizing bacteria also can be used for the microbial leaching of mining wastes and low-grade ores. The dissolution of metal from slab is a slow process, but appropriate design of the process would reduce the long-term pollution effects of ground water contamination [73].

Stage of Development

A pilot-scale industrial operation is in use.

Implications for Energy Consumption

These are not clear at present.

URANIUM MINING SITE RESTORATION AND WASTEWATER TREATMENT BY BACTERIAL LEACHING

Description

Microorganisms have been used as an alternative method of leaching ores. Presently they are also being used to restore mine and mill effluents, and current research indicates that some bacteria could play an important role in the environmental restoration of in situ uranium leaching sites. Two such applications are summarized below:

Restoration of In Situ Uranium Mines

In situ chemical mining of uranium using leaching agents such as ammonium carbonate and bicarbonate with hydrogen peroxide has flourished in the mining regions of South Texas and Wyoming. The ammonium ion, however, does have a tendency to adsorb on clays in the mineral formation, allowing ground water passing through the formation to mobilize it. This creates a potential water quality problem when the ammonium is converted to nitrite and nitrate, substances that are potentially hazardous to human health and the environment. Laws in the areas mentioned above require that the water at the mine location be restored to baseline ground water quality and the mine site be returned to its original state after chemical mining has ceased [74].

Bacteria recently have received considerable attention as possible agents for restoration of in situ mine operations leached with ammonium carbonate solutions. *Nitrosomonas* and *Nitrobacter* are two genera specifically considered for this function. Here, *Nitrosomonas* oxidizes ammonium to nitrite and *Nitrobactor* oxidizes nitrite to nitrate. Other denitrifying bacteria will reduce nitrate to nitrous oxide or ammonium ion. Following are the reactions

$$NH_4^+ + \frac{3}{2}O_2 \xrightarrow{\text{Nitrosomonas}} 2H^+ + H_2O + NO_2^-$$

$$NO_2 + \frac{1}{2}O_2 \xrightarrow{\text{Nitrobacter}} NO_3$$

In laboratory experiments, researchers have demonstrated that biological nitrification can occur in a column experiment using a uranium core and simulated ground water. More research is needed to critically examine the introduction of the nitrifying bacteria into the formation, the retention of viability of the bacteria in situ, the continued activity of the bacteria, their ability to oxidize adsorbed ammonium ion, and reduction in permeability due to bacterial clogging.

Restoration of Mining and Milling Waters

Restoration of mining and milling water has been accomplished successfully at several mining ventures. The process includes use of algae and higher aquatic plants to remove both soluble and particulate heavy metals (e.g., lead) from the mill tailings [75]. In this operation, heavy particulates are settled on a tailing pond, and the effluent from the pond passes through a series of shallow meanders containing plant growth. The analyses to date indicate that these algae and aquatic vegetation effectively accumulate or entrap the heavy metals. Vegetation identified to function in this waste water treatment system include *Cladophora, Rhizoclonium, Hydrodictyon, Spirogyra, Potamogeton* and *Oscillatoria*. Use of microorganisms in mine-associated activities do present some problems. Microbial fouling is an obvious problem for which causes and answers are not currently evident. It has been suggested that use of antimicrobial agents will prevent such clogging, but antimicrobial agents must be inexpensive and environmentally acceptable.

Stage of Development

Both laboratory-scale research and development and a pilot-scale operation are underway.

Implications for Energy Consumption

There is the potential for reduced energy consumption if compared with conventional tertiary treatment methods.

CHAPTER 5

CHEMICALS

This chapter presents technical descriptions related to the production of chemicals.

PRODUCTION OF ETHANOL FROM STARCH

Description

Traditionally, starchy raw materials are hydrolyzed by barley malt enzymes to fermentable sugars for subsequent fermentation to ethanol. Since 1960 microbial enzymes have been developed that have the potential for malt replacement in ethanol production. In the United States, Miles Laboratories has developed a thermostable α-amylase enzyme called TAKA-THERM® for starch liquefaction, and DIAZYME L-100®, a glucoamylase for starch saccharification, into fermentable glucose. The advantages in using these microbial enzymes, according to the manufacturer [76], include raw materials savings, potential increased ethanol yields, standardized enzyme activities to assure maximum yields, improved thermostability of TAKA-THERM during liquefaction, and greater stability of DIAZYME L-100 at lower pH values.

TAKA-THERM is a liquid bacterial α-amylase of *Bacillus licheniformis* var. TAKA-THERM converts starch, amylose and amylopectin to soluble dextrins and small quantities of glucose and maltose. It is thermostable at temperatures above 90°C. DIAZYME L-100 is a liquid glycoamylase derived from a selected strain of *Aspergillus niger* var. It is capable of achieving a complete breakdown of starch into fermentable sugar beginning at the nonreducing end of the starch chain.

Basic steps in alcohol production include the following [76]:

1. Milling   The starch source (usually corn, wheat, rye, sorghum or barley) is ground into a fine meal (12-16 mesh). This exposes the

		starch granules and permits suspension and dispersion in the following step.
2.	Slurrying	Water is added to the meal to form a mash (typically 15–25 gal (53–95 liters) of water per 56 lb (25.4 kg) bushel of grain). The mash is adjusted to pH 6.0–6.5 and the liquefying enzyme, TAKA-THERM, is added.
3.	Liquefaction	The mash is heated to gelatinize the starch (170–200°F, 78°–93°C) and to render it susceptible to enzyme breakdown. During this cooking step, TAKA-THERM rapidly reduces the viscosity of the mash and the starch is converted into soluble, high-molecular-weight carbohydrates called dextrins.
4.	Conversion	The mash is cooled to conversion temperature (135–140°F, 57–60°C). The pH is adjusted to 4.0–4.5 and a conversion enzyme, DIAZYME L-100, is added. The mash is held long enough to permit some of the dextrins to be converted into fermentable sugars. It is then cooled further to the fermentation temperature (85–90°F, 57–60°C). Water can be added to aid cooling to a final volume of about 33 gallons (24.7 liters) per 56 lb (25.4 kg) bushel of grain.
5.	Fermentation	Distiller's yeast is added and the mash is held 48–120 hours. During this time the conversion enzyme continues to break down dextrins to fermentable sugar while the yeast converts sugars to alcohol and carbon dioxide. The fermentation process produces heat so that cooling is necessary to maintain yeast survival.
6.	Distillation	The fermented mash is heated to vaporize the alcohol, and the vapors are collected and cooled to condense the alcohol back to a liquid. The residue contains the residual grain, spent yeast and water. The residual grain and spent yeast are generally used in animal feeds. The water usually is recycled.

Stage of Development

Enzymes are available commercially from Miles Laboratories, Elkhart, Indiana. The enzyme is used by various alcohol producers to make up to 10,000 gallons (190 proof) per year. The alcohol fuel kit is available from Economy Alcohol Fuel Supplies, Dayton, Ohio.

Implications for Energy Consumption

In this process, 20,000 Btu/gal 190 proof alcohol is required, while the energy content of the alcohol is about 85,000 Btu/gal. In other words, the energy output is four times the energy input [77].

INDUSTRIAL ALCOHOL PRODUCTION USING WHEY AND GRAIN WASTES

Description

Lactose in whey can be fermented by lactose producing *Kluyveromyces* or other genera of yeast. These yeast (especially *K. fragilis*) can reduce 5%

lactose to 0.1–0.2% in 24 hours. Table 5-1 gives the fermentation efficiency of selected organisms in whey permeate. These organisms are able to produce approximately 2% alcohol within 12 hours. The process described here uses a combination of whey and grain to produce industrial alcohol. The schematic of such a process is given in Figure 5-1.

In traditional grain fermentation, water is added to grain to create a mash, which is then cooked and treated with enzymes to convert starch to glucose. This cooked starch is then inoculated with the yeast *S. cerevisiae* and fermented at 30°C for approximately 48–60 hours. In the process described here, protein from whey is removed by ultrafiltration and the resultant permeate is added to the grain to replace all or some of the water used to create a mash. This provides protein and lactose from whey and also reduces the amount of grain needed by introducing additional fermentable sugars. After blending, the raw mash is autoclaved to sterilize mash and gelatinize starch. Starch converting enzymes α amylase and glucoamylase are blended into the gelatinized mass. The fermentation is allowed to progress at 28°C. At the end of 24 hours fermentation 9.5% alcohol is produced, while at the end of 36 hours the concentration of ethanol is increased by only about 1%.

Whey provides significant levels of fermentable sugars and the grain (corn) content of the mixed mash can be reduced correspondingly. Figure 5-2 shows alcohol production using 20% less grain than normal. Since 10% alcohol is produced within 36 hours of operation, there is substantial reduction in time and energy compared to conventional fermentation, which requires 48–60 hours. This process offers economic, as well as energy reduction, incentives.

Stage of Development

Laboratory-scale research and development is in operation.

Implications for Energy Consumption

There is a reduction in the time required for fermentation by using whey/grain mixtures.

Table 5-1. Fermentation Efficiency of Selected Organisms in Whey Permeate [78][a]

	% Residual Lactose		% Ethanol	
	12 hr	24 hr	12 hr	24 hr
K. fragilis	0.21	0.10	2.0	2.0
K. fragilis + *S. cerevisiae*	0.25	0.12	2.0	2.0

[a]The whey permeate contained 5.1% lactose initially.

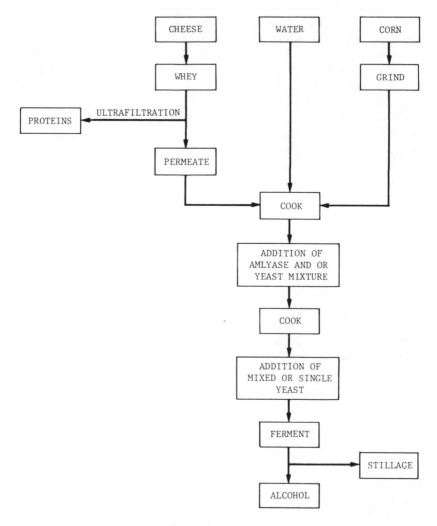

Figure 5-1. Schematic of combining whey and grain to produce industrial alcohol [78].

SIMULTANEOUS PRODUCTION OF ETHANOL AND SINGLE-CELL PROTEIN FROM CELLULOSIC WASTE

Description

This summary describes an integrated processing scheme for the conversion of a cellulosic waste, such as newsprint, to sugar by enzymatic hydrolysis,

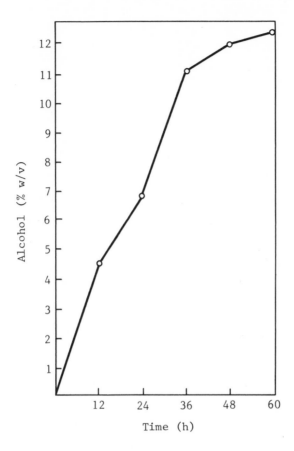

Figure 5-2.   Alcohol production in a medium in which the water is replaced by undi-
luted whey and the concentration of the grain is 15% less than normally used [78].

followed by conversion to ethanol and yeast (SCP) by fermentation. This
process is not yet optimized since consideration is not given to recovery of
the hemicellulose sugars or to more sophisticated pretreatments to remove
or breakdown lignin prior to hydrolysis. The process offers an advantage of
producing ethanol and SCP simultaneously, and has the capacity to generate
steam and power, thus achieving a net thermal efficiency of 19%. Figure 5-3
is a schematic flow diagram for the processing scheme. The primary plant
feed consists of 885 ton/day of newsprint containing 6% moisture. By means
of moderate shredding and hammer milling, newsprint is reduced to approximately 20 mesh (size of shredded paper). This shredded material is then used
to make an aqueous suspension, which can be pumped, agitated and filtered.

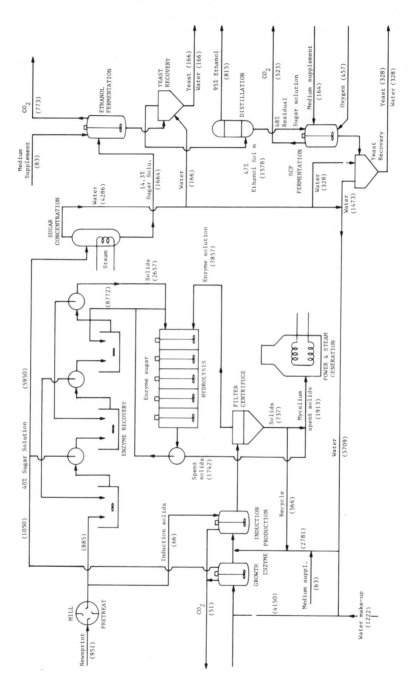

Figure 5-3.   Material balance flow diagram for integrated processing scheme [79].

A total of 66 ton/day of feed material is diverted to the first enzyme induction fermenter after sterilization with steam.

Hydrolysis is conducted over a 40-hour period at 45°C at a solid:liquid ratio of $\frac{1}{20}$ (w/w) based on inputs to the hydrolyzer. The hydrolyzer consists of five agitated cylindrical concrete digestors of the type used for solid waste treatment in sanitary engineering. The sugars from the hydrolyzer are concentrated and distributed to ethanol fermenters. The ethanol solution from the fermenter plus the aqueous stream containing the alcohol are concentrated by distillation to produce 81.5 ton/day of 95% ethanol. Yeast recovery is achieved by centrifugation and spray-drying, with an SCP yield of 32.8 ton/day of torula yeast (SCP).

Figure 5-4 shows the flow of energy among the various processing operations. At net thermal efficiency of 19% is achieved when 108.5×10^6 Btu/hr energy is produced from an input of 571×10^6 Btu/hr. With further research in the enzymatic hydrolysis process and development of more effective enzyme systems and enzyme recovery, the energy efficiency of the process can be improved.

Stage of Development

Laboratory-scale research and development is in operation.

Implications for Energy Consumption

19% thermal efficiency is achieved by integrating ethanol production with SCP production. This is desirable since the fermentation process offers two end products concomitantly, instead of separate processes, one for ethanol production and the other for SCP production.

USE OF WHOLE-CELL CATALYSIS FOR CHEMICAL MANUFACTURING

Description

Use of enzyme preparations to produce ethanol, methanol, propylene oxide and other organic chemicals is a well-accepted practice in the chemical manufacturing industry. A novel option of using whole bacterial cells as catalysts instead of prepared enzymes offers a number of potential advantages: (1) complete conversion of the substrate to the product at rates comparable to those observed when enzyme preparations are used; (2) avoidance

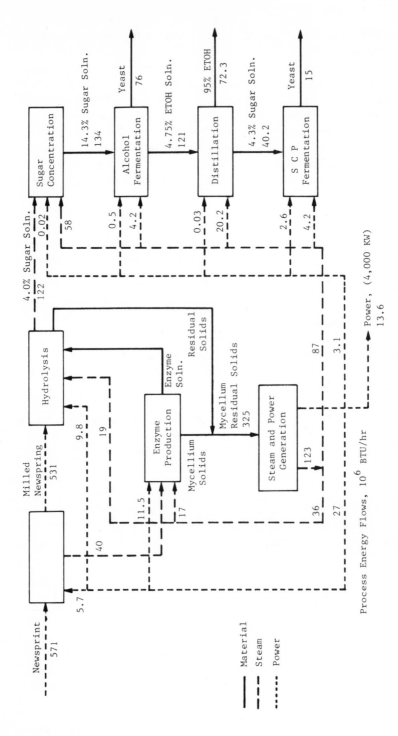

Figure 5-4.   Energy balance flow diagram for integrated processing scheme [79].

of the step of enzyme preparation from bacterial cells resulting in cost reductions; and (3) protection by the cell wall and membrane, allowing microbial enzymes to remain stable and active over long periods of time.

The use of microbial cells as catalysts has been demonstrated in the laboratory with *Methylococcus capsulatus,* inserting a single atom of oxygen catalytically into methane. The problem encountered in undertaking such catalysis with conventional chemical enzyme techniques is that methane is oxidized to methanol and the methanol produced is immediately worked on by methanol dehydrogenase; thus, methanol does not accumulate. In the present process, the microbial cells are used as specific chemical catalysts. A stream of propylene, methanol and oxygen is passed over the immobilized white cell system to produce propylene oxide. It has been noted that immobilized cells are not as active as free cells. However, their half-life is usually a lot longer and they are more stable with temperature. Such an enzyme system can be operated at only 45°C and at atmospheric pressure unlike the conventional chemical enzymes, which require high temperatures and pressures. Researchers have tested *Methylococcus capsulatus's* catalytic potential on other substrates such as ethylene, ethane, cyclohexane, benzene, toluene, methanol, styrene, and pyridine.

Stage of Development

Laboratory-scale research and development is underway.

Implications for Energy Consumption

These are not clear at present.

FOOD PROCESSING WASTES AS A SUBSTRATE FOR PRODUCTION OF INDUSTRIAL CHEMICALS

Description

In the food processing industry, there is a recognized need for methods that would convert plant effluents into useful chemicals. This is due to two facts:

1. Certain carbohydrates, proteins and biopolymers occur widely in many different wastes.
2. Various organisms and purified enzymes can stabilize many organic wastes and may produce useful products such as fuel, animal feed and chemicals.

A major type of agricultural residue (solid waste) with potential for enzymatic conversion to useful products is the cellulose contained in fruits and vegetables, and, hence, in the residual of processed foods. The long-term possibilities for bioconversion of food processing wastes are as follows:

1.   Starch → invert sugar (glucose and fructose)
2.   Biologically degradable streams $\dfrac{\text{anaerobic}}{\text{digestion}}$ methane
3.   Cellulose/chitin/fermentable sugars → alcohol/single-cell protein

In the short run, however, enzymatic processes demonstrate the greatest potential. With these processes, the true waste from vegetable and fruit processing is only 20% of the total residual material. The other 80% is utilized as animal feed or by-products.

The high variability in waste characteristics makes enzymatic conversion difficult. Nevertheless, the processes listed below involving microorganisms or enzymes have been developed up to the demonstration stage:

• Conversion of cellulose to glucose
• Conversion of glucose to alcohol
• Conversion of glucose to protein
• Conversion of starch to invert sugar
• Methane production by anaerobic digestion

Each of these processes is described briefly below. Detailed process information can be obtained from the literature cited.

Conversion of Cellulose to Glucose

The process described below is currently being developed by the Lawrence Berkeley Laboratory, Berkeley, California [79]. The cellulose hydrolysis reactions are catalyzed by a system of enzymes produced by a variety of fungi and bacteria. During the enzymatic hydrolysis, cellulosic waste is slurried and pumped into a fermenter. The fungus secretes the cellulose enzyme system using the portion of cellulosic waste as a substrate and some recycled hydrolysate (product of hydrolysis) as a carbon and energy source. The enzyme system thus produced is free of fungus cells and unreacted waste since unreacted portions are recycled. The enzyme system and cellulosic waste is mixed in a five-stage hydrolyzer for 40 hours. The product sugar serves as a starting point for crystallization and recovery or can be used as a feedstock for microbial conversion to SCP, alcohol or other chemicals. The undigested waste can be used to fuel the plant.

This process is not applicable to the waste for an individual farmer or food processor, but would require a centralized deposit site.

Conversion of Glucose to Alcohol

This process has the potential of microbially converging glucose to useful products such as ethanol, acetone, butanol, acetic acid, lactic acid and protein. This technique has been used in wine-making for thousands of years. During and following World War II, industrial alcohol production from molasses and grain was an alternative means of alcohol production. Briefly, ethanol is produced from glucose by yeasts, using the biochemical pathways of glycolysis and conversion of pyruvate to ethyl alcohol.

Conversion of Glucose to Protein

Among the processes described, production of protein from glucose conversion by yeast is the farthest advanced toward commercialization [80]. The advantages of such processes lie in the fact that yeast grows rapidly, and the fermenters are closed systems, hence reducing fluctuation in pressure, temperature, oxygen availability and other parameters. Briefly, the production of SCP involves production of cell mass under aerobic conditions. This requires the transfer of oxygen into solution by agitation. The production of SCP has the potential for generating large amounts of heat, which is due to yeast metabolism. This excess heat must be removed to maintain temperature control.

Conversion of Starch to Invert Sugar

Invert sugar (glucose and fructose) either can be produced by hydrolysis of sucrose or by saccharification of starch and isomerization of glucose. The starch can be made available as waste starch streams from food processing wastes following a step that concentrates starch components. The conversion of starch to glucose by application of immobilized enzyme technology is the largest single application of enzymes today [81]. A production plant has been installed by Corning at Iowa State University to produce dextrose from cornstarch, which is not waste material. It has been shown at the Iowa State University plant that such a system can reduce reaction time from several days to minutes. The extent of reaction is controlled by temperature and by the feed rate, or time of reaction. Even though this system has been developed for cornstarch as a substrate, with some modifications it can be adapted for using starch from food processing wastes.

Methane Production by Anaerobic Digestion

This process offers a two-step approach for converting food processing wastes into methane. The first step of acid formation involves oxidation of wastes to produce acetic acid and to generate ammonia, carbon dioxide and hydrogen sulfide. This is done by common anaerobic facultative bacteria and proceeds at a rapid rate; typical time required for this step may be several hours to a day. The second step involves the reduction of carbon dioxide and cleavage of the acetic acid. Several methanobacteria have been identified as catalysts acting symbiotically. During this step, bacterial growth and their sensitivity to environmental conditions are the rate-limiting factors. At optimized conditions the residence time is suggested to be 5–15 days when municipal wastes are used as a substrate.

The use of food processing waste treatment for methane production is a technology with a potential for reduced energy consumption since this process does not require expensive and energy-inefficient pretreatment steps, e.g., milling or separation.

Food processing wastes are particularly well suited for anaerobic digestion since they contain sugars, starch, proteins and organic acids.

Stage of Development

Laboratory-scale research and development and commercial (conversion of glucose to protein) production are underway.

Implications for Energy Consumption

There is the potential for a reduction in energy consumption when compared with food processing and municipal waste treatment requirements.

BIOPOLYMERS FROM MUNICIPAL WASTE

Description

It has been demonstrated that mixed cultures of microorganisms in activated sludge from a waste treatment plant, when aerated with low concentrations of methanol, produce a highly viscous crude product containing glycan polymer [82]. During treatment of the activated sludge from an industrial corn products firm, heteropolysaccharides were produced only if the acid formed was neutralized repeatedly during the waste treatment. This heteropolysaccharide, the end product of fermentation, contains glycan polymer as one of the constituents.

The process described here includes an automatic continuous addition of methanol which, for large-scale operations, might provide an efficient utilization of substrate, i.e., municipal waste. The process involves use of 20-liter-capacity fermenters with 10 liters of activated sludge and 1% v/v methanol. The incubation is carried out at 25°C with 2% additional methanol added after 2 days. Following 7 days of fermentation, products from the fermenters are tested for viscosity and methanol utilization. During processing, biopolymer formation, as judged by increased viscosity, tapers off and methanol concentration becomes low. Continuous addition of methanol prolongs the process over the 7-day fermentation period. The viscosity exhibited by the biopolymers in admixture with activated sludge is suspected to be due to the composition of the biopolymer, the amount of polymer formed, the length of polymer chains, the physical aggregation of the chains, or a combination of these. At the time of maximum viscosity, the biopolymer apparently may be stabilized by addition of methanol to maintain low concentration.

The activated sludge permits a nonsterile fermentation with a self-selected mixed culture. The advantage of this process is that the starting material is constantly available during the fermentation period, while the process requires only the addition of methanol and air. The end product can be pumped out of the fermenters easily, and a small portion of the fermented product can be reused as a backseed.

This crude product (heteropolysaccharide) suspension, when sprayed on a surface following evaporation of the water, leaves a friable film that is not easily redispersed or displaced by water. This property has been used by scientists to prepare mixtures of fertilizers or pesticides in which the viscous product acts as a slow release agent [83].

Stage of Development

A pilot plant is in operation.

Implications for Energy Consumption

Resource conservation is indicated.

PULP AND PAPER INDUSTRY WASTE AS A SUBSTRATE FOR LACTIC ACID PRODUCTION

Description

Inclusion of a fermentation component in wastewater treatment has the advantage of having a higher throughput rate than a conventional wastewater treatment process. For example, most conventional wastewater treatment processes have loading rates of about 100–200 lb (45.4–89.8 kg) of chemical oxygen demand (COD) per 1000 ft^3 (305 m^3) of process tank volume per day. If a modern tower fermentation is used to produce 5% ethanol, the loading rate would be in excess of 16,000 lb/1000 ft^3 of overall tank volume per day. According to researchers, this represents a considerable savings in treatment facility capital costs and overall volume.

In the laboratory-scale version of this process, wastes from the pulp, paper and fiberboard industries are used as substrates. The waste is diluted prior to use as a fermentation feed and buffered with lime to increase the pH to conditions favorable to lactic acid production. A fixed-film unit is seeded with a mixed commercial bacterial/yeast culture containing lactobacilli and lactose-fermenting yeasts derived from a commercial kefir culture. A fixed-film unit has a coating of gelatin cross-linked with 5% glutaraldehyde, which facilitates microorganism attachment. The mixed cultures are stabilized by using whey as a substrate for several weeks prior to operation using wood molasses (wastes from pulp, paper and fiberboard industries). The buffering achieved by adding lime influences the amount of lactic acid produced while enzyme pretreatment has no effect on the lactic acid yield. The results are shown in Table 5-2. The disadvantage of buffering is that scaling might occur on the full-scale fixed-film system. Secondly, the buffer addition might require the use of process control equipment. For concentrated wastewater, fermentation with distillation or other physical or chemical methods of product recovery can be an attractive alternative, especially when the recovered value of the product is considered. The process described below can be applied to concentrated wastewaters from the dairy, wood fiber, citrus and canning industries. It has been suggested that the process described here consisting of small-volume reactors and simplicity of operation will provide a useful method of by-product recovery from high-carbohydrate wastes. In doing so, it has a potential to offer low energy consumption with better product yield.

Table 5-2. Lactic Acid Production from Masonite Wood
Molasses Using the "Fixed-Film System" [84]

Run	Feed Carbohydrate (%)	Enzyme Pretreatment	Time (days)	Residual Reducing Sugar (%)	Lime Buffer (meq)	Latic Acid (g/l)
A	4		1	0.25	75	13
B	4	A. niger cellulase	1		75	14
C	4	A. niger cellulase (10X)	1	1	75	14
D	4	Diastase + pentosanase	1		75	13
E	4	T. viride cellulase	1	0.75	75	13
F	4	T. viride cellulase	3	0.75	75	11
G	8			1	125	25
H	8			1	125	26
I	8			0.75	250	50
J	6(fresh)				250	25

Stage of Development

Laboratory-scale research and development is underway.

Implications for Energy Consumption

This is a potentially energy-efficient process.

PRODUCTION OF ETHANOL AND ACETIC ACID FROM CELLULOSIC WASTE

Description

Biomass utilization for energy production is the enzyme-catalyzed hydrolysis of cellulose and hemicellulose to low-molecular-weight components that can serve as substrates for fermentation to fuels or chemicals. Here, a technique is described in which cellulose degradation, cellulose hydrolysis and fermentation can be carried out simultaneously in a single operation. This process eliminates the need for separate enzyme production and preparation hydrolysis reactors, thus reducing energy requirements for two separate operations. Furthermore, in this process cellulose can be converted directly to useful products such as ethanol and acetic acid along with a simultaneous accumulation of fermentable sugars. Figure 5-5 shows a hypothetical biochemical pathway of cellulose degradation and end product formation.

Clostridium thermocellum, an anaerobic and thermophilic bacterium, is used in the laboratory-based technique. Bacterial cultures are grown under anaerobic conditions in a specifically prescribed medium [85], and Solka Floc SW 40 or ball-milled corn residue or cellobiose are used as a cellulose substrate in varying amounts.

A bench-scale fermenter (2–5 liter) is used with agitation speed maintained at 100 rpm. Anaerobiosis is achieved and maintained by introducing N_2 or CO_2 during the fermentation. The age of the bacterial culture is critical, and 18-30 hours of culture prior to use has been shown to provide an actively growing culture. The pH is maintained at 7.1 while temperature is controlled at 60°C.

Table 5-3 shows specific rates of product formation by *Clostridium thermocellum.* Corn residue is shown to exhibit the highest rates of cellulose degradation and product formation. The concentration of end products can be increased by implementing batch-fed fermentation where substrate is

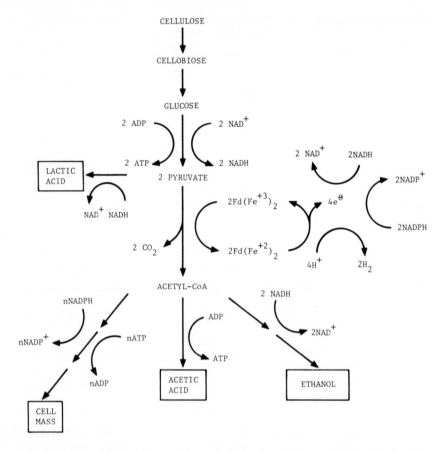

Figure 5-5. Hypothetical biochemical pathway of cellulose degradation and end product formation [85].

added at intervals during the fermentation period of 60 hours. A total of 47.5 g cellulose is fed and 16 g remains at the end of fermentation, representing 66% degradation, while 4 g/l each of ethanol and acetic acid are produced. A drawback of this process is that ethanol and acetic acid concentrations could be inhibitive to the growth of *Clostridium*. This problem can be resolved, however, by using *Clostridium* cultures grown on media containing concentrations of ethanol and acetic acid. This laboratory-developed mutant, *Clostridium thermocellum,* is suspected of converting cellulose to ethanol at 85% of theoretical yield [85].

Table 5-3. Specific Rate of Product Formation by *Clostridium thermocellum*
on Different Cellulose Substrates and Cellobiose [85]

Time of Fermentation (hr)	Reducing Sugar (g/g cell-hr)	Ethanol (g/g cell-hr)	Acetic Acid (g/g cell-hr)
Corn Residue			
3	0.53	0.23	0.34
4	1.03	0.20	0.32
6	0.98	0.19	0.36
8	0.96	0.23	0.38
10	0.52	0.23	0.16
Solka Floc			
4	0.35	0.11	0.07
7	0.36	0.13	0.05
10	–	0.16	0.06
Cellobiose			
4		0.26	0.16
6		0.27	0.20
8		0.20	0.12
10		0.12	0.10
12		0	0.11

This simultaneous production process has the potential of reducing both the capital investment and operating costs when compared to the two-step conventional process. Figure 5-6 summarizes the block diagram for a hypothetical example. This process offers an overall yield of ethanol at 35% of the solids degraded. More research is necessary toward process improvement.

Stage of Development

Laboratory-scale research and development is in operation.

Implications for Energy Consumption

The simultaneous production of ethanol and acetic acid will consume less energy than conventional processes of separate production.

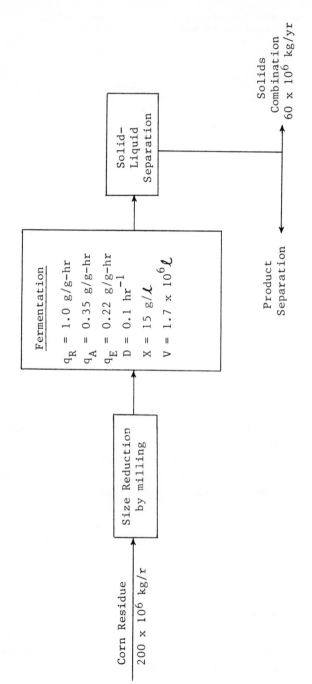

A facility handling 200×10^8 kg of corn residue fed per year with a fermentable polysaccharide content of 70% on a dry weight basis. Ethanol: 16×10^6 kg/yr; yield = 0.11 g/g. Acetic acid: 25×10^6 kg/yr; yield = 0.18 g/g. Reducing sugar: 71×10^6 kg/yr; yield = 0.51 g/g.

Figure 5-6.   Block diagram of a direct cellulose fermentation process scheme [85].

CHAPTER 6

METALS RECOVERY

URANIUM EXTRACTION BY BACTERIAL LEACHING

Description

The term "bacterial leaching" is used to describe the solubilization of metals from their ores in an operation that is relatively crude consisting of the percolation of acidified water through heaps of broken, low-grade ore. Within such heaps, bacterial activity results in mineral sulfide oxidation and release of metals [86].

The process of bacterial leaching of metals such as copper and uranium from their ores is based principally on the activity of *Thiobacillus ferrooxidans*. Some metal extraction may be attributable to other organisms, which stimulate the activity of *T. ferrooxidans*. *Thiobacillus* has been shown to be inhibited by the presence of mineral flotation processing reagents.

In hydrometallurgical processes, bacterial leaching is an important, emerging technique for the extraction of metals. The conventional physical leaching methods include dump, heap, vat and in situ leaching, where the leaching solutions are usually introduced on the top surface of a dump, heap or vat by flooding. In physical leaching it is necessary that leach solutions reach all metal-bearing materials, that an adequate amount of air be available, and that the temperature of the leach solution be maintained at a constant level. These conventional methods are often costly and present serious pollution problems. By contrast, the advantages of the bacterial leaching process are as follows: (1) the relative absence of land and water pollution in the process; (2) the ability to mine increasingly lower-grade ores that cannot be processed economically by conventional mining, smelting and refining operations; (3) ease of operation; and (4) lower capital costs as compared with conventional processing. Sulfuric acid produced by *T. ferrooxidans* reacts on metal ores resulting in solubilization of metals from ores. Thus, uranium may be extracted into acid solution as follows:

$$UO_2 + Fe_2(SO_4)_3 \rightarrow UO_2SO_4 + 2\ FeSO_4$$

$$UO_3 + H_2SO_4 \rightarrow UO_2SO_4 + H_2O$$

Physical leaching of uranium has been successful on a commercial scale for a number of years. Uranium deposits, particularly those in sands that have controllable permeability, are most amenable to solution mining.

The role of bacteria in uranium leaching consists of two processes: ferric ion generation and uranium leaching. Successful uranium extraction in a two-stage system has been in operation on a production scale in Canada [86]. More recently, the iron-oxidizing bacterium *Thiobacillus ferrooxidans* has been used successfully for bacterial leaching of uranium ores by Agnew Lake Mines Ltd. in Northern Ontario, Canada. At this mine, the primary uranium ore is uranothorite $((Th,U)SiO_4)$, which is readily soluble in weak acid. The operation consists of blasting underground stops to break the ore. The "swell" due to the explosion is leached on the surface with acidic ferric sulfate generated by *T. ferrooxidans.* The underground ore is then leached by percolating the solution through the ore. The acid ferric lixiviant is pumped into the ore material and withdrawn to the surface after reaction and solubilization of metals by the ferric iron. This type of in situ mining by bacterial leaching is an attractive option due to the reduced environmental impact over conventional mining, the feasible extraction of ore from deeper deposits and the favorable economics of mining low-value mineral reserves.

Genetic selection for organisms that have increased leaching rates for specific minerals and greater tolerance to toxic metals is desirable, but the present stage of knowledge is not advanced enough to explore and study genetic manipulation for this kind of strain selection. Secondly, there is a lack of basic descriptive genetic information about the thiobacilli, and even fewer basic data on other potentially important leaching organisms, such as *Sulfolobas* and *T. organoporus.*

Stage of Development

Commercial production is carried out by Agnew Lake Mine Ltd., Ontario, Canada. (NOTE: According to Dr. Tuovinen of Agnew Lake Mine, Ltd., the uranium extraction by bacterial leaching at this mine was discontinued recently. At present, details about the shutdown are not available [73].)

Implications for Energy Consumption

These are not clear at present.

COPPER EXTRACTION BY BACTERIAL COLUMN LEACHING (LABORATORY SCALE)

Description

Traditionally, metal extraction is achieved by leaching chemical solution through ore dumps or heaps. Bacterial leaching of metals from sulfur minerals has been studied for a number of years. In particular, microorganisms that grow at high temperatures and at low pH, designated as "ferrolobus" and *Sulfolobus acidocaldarius,* are being used in the laboratory to oxidize reduced iron and sulfur compounds, respectively. These thermophilic bacteria leach metals from porphyry copper samples in which chalcopyrite ($CuFeS_2$) is the primary mineral.

For many years copper leaching in field mining has been achieved successfully by using *Thiobacillus ferrooxidans.* For their continued growth and activity, these bacteria need carbon dioxide and small amounts of ammonium ion, phosphate and trace elements. The sulfuric acid produced by *T. ferrooxidans* reacts on copper metal ores resulting in solubilization of metals from ores [86].

$$CuFeS_2 + 2 Fe_2(SO_4)_3 + 2 H_2O + 3 O_2 \rightarrow CuSO_4 + 5 FeSO_4 + 2H_2SO_4$$

$$Cu_2S + Fe_2(SO_4)_3 \rightarrow 2 CuSO_4 + 4 FeSO_4 + S$$

The laboratory-scale method described here uses *S. acidocaldarius,* a thermophilic bacterium, and leaching is conducted at an incubation temperature of 60°C.

The leaching solution consists of salts diluted with distilled water and pH adjusted to 2.5 with sulfuric acid. The bacterial culture contains growth medium supplemented with yeast extract and ferrous ion or sulfur. The leaching is allowed to continue for 60 days in Erlenmeyer flasks. At the end of 60 days, Ferrolobus leached 51% of the copper from chalcopyrite concentrate containing 29% copper. Addition of yeast extract enhanced growth of the organisms, which could directly attack chalcopyrite. The impact of adding ferrous ion and sulfur ion as an energy source has not yet been determined. Column leaching is another method by which copper ores are leached in the presence of thermophilic bacteria. Figure 6-1 shows a

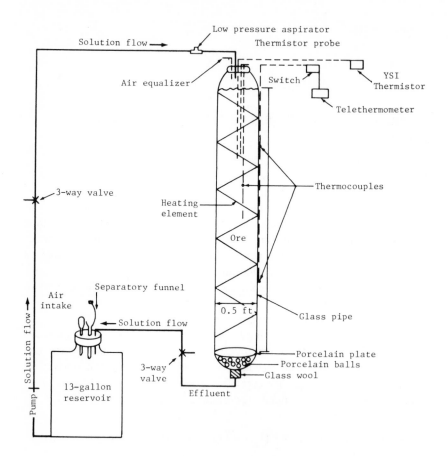

Figure 6-1.   Schematic view of a column leaching system [87].

schematic of the column leaching system. In this system, mixed cultures of the thermophilic bacteria leached 38% of the copper from −3 to +100 mesh chalcopyrite in 161 days. The amount of metal recovered from ores depends on the mesh size of ore particles.

The results of this laboratory study indicate that thermophilic bacteria could be useful in in situ mining where temperatures could be maintained for optimum growth of the organisms. These organisms also could be of value to vat leaching systems and dump leaching operations in which exothermic chemical reactions increase the temperature sufficiently for thermophilic bacterial growth. Figure 6-2 illustrates the role of microorganisms in leaching operations.

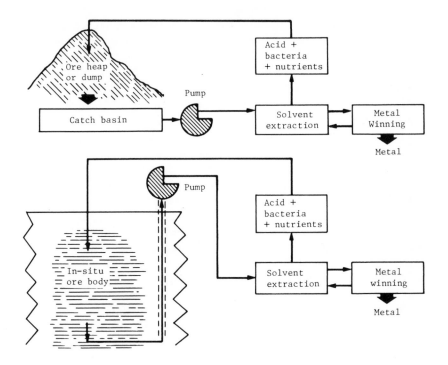

Figure 6-2. Role of microorganisms in leaching operations is shown here. Leaching of ore heap or dump is at top; leaching of in-place ore body (in situ mining) is illustrated at bottom [88].

Stage of Development

Laboratory-scale research and development is in operation.

Implications for Energy Consumption

There are no apparent impacts on energy consumption.

USE OF BACTERIA IN METALS REMOVAL FROM LANDFILL LEACHATES

Description

Trace metals in leachates from municipal landfill operations pose an environmental problem resulting in the uncontrolled release of toxic leachates to receiving streams and ground waters. Aerobic digestion has been shown to be an effective method of removing the trace metals such as aluminum, cadmium, calcium, chromium, iron, manganese, zinc, lead, nickel, etc. Other physical-chemical treatments, singularly or in combination also have been shown to be effective to some extent.

Here, a laboratory-scale process of anaerobic digestion of leachates is outlined. Bench-scale anaerobic digesters are used operating in the mesophilic range ($29°$-$38°C$) with a capacity of 14 liters. The digester startup is achieved by introducing 7 liters of anaerobically digested, domestic waste water sludge and operating the digestor for 24 hours to ensure that anaerobic conditions prevail. The digestor is equipped with a mechanical mixer for continuous stirring and a thermostatically controlled heater maintaining temperature at $34°\pm1°C$. Following 24 hours of operation, leachate is added at the rate of 0.5 liter/day for a period of 2 weeks. By then the digestor is filled to its capacity and allowed to become fully stabilized. During the 2-week period, the digestor stability is determined by monitoring pH, alkalinity, suspended solids, volatile suspended solids, volatile acids, gas production rate and gas composition.

The pH is adjusted by adding $Ca(OH)_2$ and nutrients in the form of diammonium phosphate to maintain an adequate BOD:nitrogen:phosphorus ratio (100:5:1). This stabilizes the digestion process.

Traditionally, trace metals are separated from mixed effluent by centrifuging at a relative centrifugal force (RCF) of 950 g. This high RCF, coupled with a relatively long centrifuging time (20 minutes), has been demonstrated to give good separation of mixed effluent metals into their dissolved and suspended fractions. The results to date [89] indicate that anaerobic digestion is an efficient metal removal process. Another method that is used to remove metal is complete settling of anaerobic leachates where metal removal efficiencies approach those of centrifugation. Table 6-1 is a summary of metal percentage in dry sludge following complete settling of anaerobic leachate. This alternative method of anaerobic leachate settling has the potential of reducing energy consumption necessary for centrifugation. Another advantage of anaerobic digestion is that the methane gas produced (50% of the total gas) can be used to operate the digestor under mesophilic temperature conditions. It has been suggested that methane production is almost certain to be well in excess of the necessary requirements for heating purposes and might, in fact, reduce energy consumption when incorporated in an energy recovery scheme. The feasibility of this is yet to be worked out.

Stage of Development

Laboratory-scale research and development is in operation.

Table 6-1. Summary of Metal Percentage Contained in Dry Sludge [89]

Parameter	Percentage
Aluminum	94.3
Barium	92.5
Cadmium	100
Calcium	30.6
Chromium	44.6
Copper	40.4
Iron	79.7
Lead	50.3
Magnesium	9.9
Manganese	68.5
Mercury	100
Nickel	86.2
Potassium	5.6
Sodium	3.7
Zinc	95.0

Implications for Energy Consumption

Metal recovery by complete settling of anaerobic leachate is an energy efficient alternative to centrifugation. Further, methane produced during anaerobic digestion can be recovered and used to maintain mesophilic temperature for the digestor.

CHAPTER 7

INSTRUMENTATION

IMMOBILIZED MICROBIAL ENZYME ELECTRODES

Description

Enzymes are known to be excellent indicators for chemical analysis and monitoring by fluorometric and electrochemical methods. The use of immobilized enzymes offers many advantages over conventional enzyme reagents. One advantage is a pH shift, i.e., by choosing the right support for immobilization, the pH optimum can be shifted to that region at which one wants to make measurements. Furthermore, immobilized enzymes are more stable, more selective and sensitive to many inhibitors, and can be reused. Bioprobes (enzyme electrodes) prepared by using immobilized enzymes are described here.

Enzyme electrodes are easy to construct when the enzyme is available either from commercial or microbial sources. The particular enzyme is immobilized by standard techniques. The immobilized enzyme is then placed around the appropriate electrode to monitor the reaction that occurs with the substrate. In a typical example of an enzyme electrode the following reaction takes place:

$$\text{Glucose} + O_2 \xrightarrow{\text{glucose oxidase}} H_2O_2 + \text{gluconic acid}$$

The enzyme electrode enables one to measure the oxygen uptake or to record the peroxide polarographically in the above reaction.

The stability of the electrode depends on the type of entrapment. The entrapment can be either physical or chemical bondage. The physically entrapped enzymes last about 3–4 weeks or 50–200 assays, while the chemically bound enzymes are good for 200–1000 assays. Enzyme electrodes have been

173

used successfully in amino acid analysis, creatinine assay, triglyceride analysis and protein measurement.

In amino acid analysis for L-arginine and 1-lysine with bioprobes, the decarboxylase enzyme used in the electrode is from *Escherichia coli*. Here the amino acid electrode is simply a base CO_2 or ammonia sensor. An enzyme layer is placed on top of this CO_2 or NH_3 membrane. The typical response time of such an electrode is of the order of 1-3 minutes, and the return to baseline is about 10 minutes.

In the assay system developed to determine serum creatinine, the enzyme electrode contains a pure creatininase from *Pseudomonas* that hydrolyzes creatinine to N-Methyl hydantoin and ammonia. The enzyme is coupled covalently onto the ammonia sensor. The bioprobe has been shown to tolerate temperatures up to 40°C.

The determination of triglycerides in plasma or serum with the present methods involves four steps in which NADH is followed spectrophotometrically and fluorometrically by the decrease in absorbance at 340 nm or fluorescence at 365, 460 nm. The main problem with this assay is that the four-step procedure requires four enzymes and three cofactors. This is expensive, requires considerable reagent preparation and is subject to "failures" that are difficult to trace [90]. An alternative approach is a simple one using fungal lipase and glycerol dehydrogenase, which catalyze the following reactions:

$$\text{Triglycerides} \xrightarrow{\text{lipase}} \text{glycerol}$$
$$\text{glycerol} + \text{NAD} \xrightarrow{\text{glycerol dehydrogenase}} \text{NADH} + \text{dihydroxyacetone}$$
$$\text{NADH} + O_2 \xrightarrow{\text{peroxidase and Mn}^{2+}} H_2O + 2\,\text{NAD}^+$$

measuring the decrease in O_2 with an O_2 electrode. The results to date have been encouraging [90].

The protein electrode being developed can be used to assay total protein in foods, e.g., soybean, peanuts. In the food preparation, enzymes are used to hydrolyze the protein to liberate L-tyrosine, which is then measured by an L-tyrosine-specific electrode. This electrode can be constructed using L-tyrosine decarboxylase from *Escherichia coli*, which catalyzes the reaction

$$\text{L-tyrosine} \xrightarrow[\text{decarboxylase}]{\text{tyrosine decarboxylase}} \text{tyramine} + CO_2$$

The enzyme is immobilized on the outer membrane of a CO_2 electrode. This electrode offers electrochemical measurements.

Stage of Development

Bioprobes are available commercially from Miles Laboratories and Technicon Inc. There is a laboratory-scale operation for triglycerides and protein electrodes.

Implications for Energy Consumption

None are apparent at present.

BACTERIAL ELECTRODE AS GAS SENSOR

Description

Commercially available L-glutamine sensors use synthetic chemicals. The bacterial electrode described here is based on the concept of a membrane electrode probe that uses intact living bacterial cells in situ to produce a highly selective and sensitive potentiometric response to the amino acid L-glutamine in aqueous standards and in human serum.

The electrode consists of a layer of whole cells of the bacterium *Sarcina flava* held at the surface of an ammonia-sensing membrane electrode with a dialysis membrane [91]. The bacteria are grown on agar slants of nutrient broth at 30°C for 3 days, then harvested and washed by centrifugation in *tris*-HCl buffer (pH 7–5) containing 0.01 mol/ℓ of $MnCl_2$ as activator.

Here, bacteria function as selective biocatalysts to convert glutamine to NH_3. This conversion produces a change in the measured potential. It is suggested that steady-state potential can be reached within 5 minutes or less. Over the linear concentration range of 10^{-4} to $10^{-2}M$ glutamine, the response slope of the electrode is −48.5 mV/concentration decade. This electrode sensitivity is suggested to be more than adequate since the normal glutamine concentration falls in the range of 4.2 to 7.6×10^4M.

The electrode is also suggested to be very selective for glutamine when tested with other amino acids such as aspargine, aspartic acid, histidine, etc. The bacterial electrode has been demonstrated to be equally responsive to glutamine concentrations in the reconstituted pooled serum (1:5 dilution).

This bacterial electrode is shown to be an effective glutamine sensor for at least two weeks following preconditioning with storage of the electrode in the buffer solution.

Bacterial electrodes of this type lay the groundwork for the development of other sensors through the appropriate combination of various bacterial strains and gas-sensing elements.

Stage of Development

Laboratory-scale research and development is in operation.

Implications for Energy Consumption

None are apparent at present.

IMMOBILIZED WHOLE CELLS AS BOD SENSORS

Description

The BOD test is one of the most widely used and important tests in the measurement of organic pollution. In practice, it requires a five-day incubation period at 20°C and demands skilled technicians to perform it. Simple and reproducible methods for estimation of 5-day BOD are being developed for pollution control. One such method uses immobilized bacterial cells in the production of a microbial electrode.

The immobilization of whole cells of hydrogen-producing bacteria, *Clostridium butyricum*, on the electrode can be accomplished as follows: the bacteria are isolated from soil and are cultured under aerobic conditions at 37°C for 9 hours in 80 ml of medium (pH 7.0) containing 1% glucose, 4% peptone, 4% yeast extract, 2% beef extract, 12.5 g/ℓ K_2HPO_4 and 0.5 g/ℓ $FeSO_4$. The cells are washed twice with oxygen-free 0.1 M phosphate buffer (pH 7.0, 5°C). The immobilization of *C. butyricum* on collagen membrane is accomplished by preparing a suspension of 1.8-g collagen fibrils and 0.6-g wet cells. This suspension is cast on a Teflon plate at 20°C. The bacterial-collagen membrane is treated with 0.1% glutaraldehyde solution for 1 minute and dried at 4°C. The bacteria-collagen membrane is used to assemble biofuel cells and microbial electrodes.

The biofuel cell uses bacterial cell immobilization techniques, which are described in detail by Karube et al. [92]. Figure 7-1 is a schematic diagram of a BOD sensor biofuel cell employing immobilized *C. butyricum*. The cell consists of anode and cathode chambers separated by an anion exchange membrane. The cathode is a carbon electrode (4 X 4 X 0.6 cm^3). The catholyte is 100 ml of 0.1 M phosphate buffer (pH 7.0). Wastewaters are introduced into an anode chamber. The current is measured by a millivolt-ammeter and the obtained signal is displayed on a recorder.

The optimum pH of the biofuel cell is 7.0 and the optimum temperature

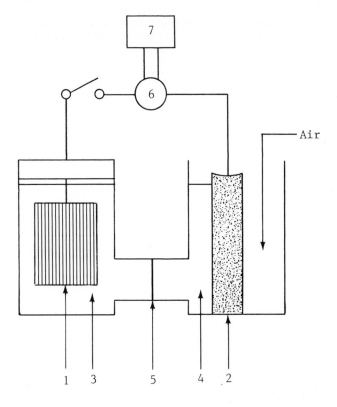

1. microbial electrode
2. carbon electrode
3. sample waste water
4. catholyte (0.1 M phosphate buffer)
5. anion exchange membrane
6. ammeter
7. recorder.

Figure 7-1.   Biofuel cell as BOD sensor [92].

required is 37°C. The biofuel cell can be applied to the estimation of the BOD of wastewaters. When the BOD (ppm) of a conventional method is compared with BOD obtained by the biofuel cell, the mean difference in values is 9.8%, as given in Table 7-1. The advantage of this biofuel cell over the conventional methods and microbial electrode is that the biofuel cell uses a single species

Table 7-1. Comparison Between the BOD Values Determined by the Conventional
Method and Those Determined by the Microbial Electrode [92]

Waste Water[a]	BOD (ppm)		
	Conventional Method	Microbial Electrode	Differences (%)
A(1/20)	29	32	10
B(1/10)	58	50	13
B(1/2000)	14	14	0
B(1/1000)	28	24	14
C(1/2000)	110	98	11
C(1/1000)	220	190	14
D(1/200)	31	35	13
D(1/100)	62	64	3

Mean difference = 9.8%

[a]A: slaughterhouse waste water; B: fermentation waste water; C: alcohol waste water; D: food factory waste water.

of bacteria, whereas other methods use many species of microorganisms obtained from soil. In spite of using only a single species, the biofuel cell is capable of giving reproducible results in the estimation of BOD.

Stage of Development

Laboratory-scale research and development is underway.

Implications of Energy Consumption

These are not clear at present.

REFERENCES

1. Messing, R. A., Ed. *Immobilized Enzymes for Industrial Reactors* (New York: Academic Press, Inc., 1975), p. 2.
2. Cape, R. A Workshop on Genetic Engineering, presented at The MITRE Corp., March 11-12, 1980.
3. Gasner, L. L. "Microorganisms for Waste Treatment," in *Microbial Technology Volume 2: Fermentation Technology*, 2nd ed., Peppler and Perlman, Eds. (New York: Academic Press, Inc., 1979), pp. 211-222.
4. Moon, N. J., and E. G. Hammond. "Oil Production by Fermentation of Lactose and the Effect of Temperature on the Fatty Acid Composition," *J. Am. Oil Chem. Soc.* 55(10): 683-688 (1978).
5. Clark, W. S. *Proc. Whey Products Conf.* Eastern Regional Research Center, Philadelphia, Pa (1977), p. 4.
6. Coton G. "The Utilization of Permeates from the Ultrafiltration of Whey and Skim Milk," address presented to the International Dairy Federation, September, 1979, Geneva, Switzerland.
7. Dohan L. A., J. L. Baret, S. Pain and P. Delalande. "Lactose Hydrolysis by Immobilized Lactose: Semi-Industrial Experience," paper presented to the Fifth Enzyme Engineering Conference, sponsored by the Engineering Foundation at Henniker, NH, July 29-August 3, 1979.
8. Moo Young, M., et al. *Biotechnol. Bioeng.* 19: 527 (1977).
9. Moo-Young, M., A. J. Daugulis, D. S. Chahal and D. G. Macdonald. "The Waterloo Process for SCP Production from Waste Biomass," *Process Biochem.* 38-40 (October 1979).
10. Richardson G. H., G. L. Hong and C. A. Ernstrom. "The Utah State University Lactic Culture System," Dept. of Nutrition and Food Science, Utah State University, Logan, UT (August 1979).
11. Crisan, E. V., and S. G. Sorensen. U.S. Patent 4,007,283, assigned to the Regents of the University of California (February 8, 1977).
12. Cheryan, M. "Application of Immobilized Proteases in the Continuous Manufacture of Cheese," *Am. Inst. Chem. Eng. Symp. Series* #172 47-52 (1978).
13. Cheryan, M., P. J. Van Wyk, N. F. Olson and T. Richardson. "Secondary Phase and Mechanism of Enzymatic Milk Coagulation," *Dairy Sci.* 58: 477-481 (1975).
14. Yang, H. Y., F. W. Bodyfelt, K. E. Berggren and P. K. Larson. "Utilization of Cheese Whey for Wine Production," *Proc. 6th Nat. Symp. of Food Processing Wastes*, U.S. Environmental Protection Agency (1975).

15. Yang, H. Y., F. W. Bodyfelt, R. E. Berggren and P. K. Larson. "Utilization of Cheese Whey for Wine Production," U.S. Environmental Protection Agency, Industrial Environmental Research Laboratory, EPA-600/2-77-106, Cincinnati, OH (June 1977).
16. Palmer, G. M. "Wine Production from Cheese Whey," U.S. EPA Industrial Environmental Research Laboratory, EPA-600/2-79-189, Cincinnati, OH (October 1979).
17. Coughlin, R. W., M. Charles and K. Julkowski. "Experimental Results from a Pilot Plant for Converting Acid Whey to Potentially Useful Products," *Am. Inst. Chem. Eng. Symp. Series* #172, 74:40-46 (1978).
18. Coughlin, R. W., and M. Charles. "Improved Utilization of Immobilized Enzymes Using Fluidized Bed Reactors," *Proc. Grantees-Users Conf. Enzyme Technol. and Renewable Resources*, NTIS Publication PB-259615 (1976).
19. Fullbrook, P. D. "Recent Advances in the Use of Enzyme in Starch Processing," *Proc. Inst. Food Sci. Technol. (U.K.).* 9(3):105-112 (1976).
20. Dworschaek W., and R. Lamm. U.S. Patent No. 3,666,628 (1972).
21. Antrim, R. L., W. Colilla and B. J. Schnyder. "Glucose Isomerase Production of High-Fructose Syrups," in *Applied Biochemistry and Bioengineering. vol. 2, Enzyme Technology.* Wingard, Katchalski-Katzir and Goldstein, Eds. (New York: Academic Press, Inc., 1979).
22. Litchfield, J. M. "Microbial Protein Production," *Bioscience* 30(6):387-396 (1980).
23. Pure Culture Products, Inc. "Product Sheet."
24. Lewis, C. W. "Energy Requirements for Single Cell Protein Production," *J. Appl. Chem. Biotechnol.* 26:586-575 (1976).
25. Shennan, J. L., and J. D. Levi. "The Growth of Yeasts on Hydrocarbons," *Prog. Ind. Microbiol.* 13:1-48 (1974).
26. Grobbelaar, J. U. "Observations on the Mass Culture of Algae as a Potential Source of Food," *S. Afr. J. Sci.* 75(3):133-136 (1979).
27. Osborne, R. J. W. "Production of Frozen Concentrated Cheese Starters by Diffusion Culture," *J. Soc. Dairy Technol.* 30:40-44 (1977).
28. Lawrence, R. C., and T. D. Thomas. "The Fermentation of Milk by Lactic Acid Bacteria," in *Symp. Soc. for General Prospects*, A. T. Bull, D. C. Ellwood and C. Ratledge, Eds. (New York: Cambridge University Press, 1979), pp. 187-219.
29. Sharpe, M. E. "Lactic Acid Bacteria in the Dairy Industry," *J. Soc. Dairy Technol.* 32(1):9-18 (1979).
30. MacBean, R. D., R. J. Hall and P. M. Linklater. "Analysis of pH-Stat Continuous Cultivation and the Stability of Mixed Fermentation in Continuous Yogurt Production," *Biotechnol. Bioeng.* 21:1517-1541 (1979).
31. Lang, F., and A. Lang. "New Methods of Acidophilus Milk Manufacture and the Use of *Lactobacillus bifidus* Bacteria in Milk Processing," *Aust. J. Dairy Technol.* 32(2):66-68 (1978).
32. Kelley, S. J., and L. G. Butler. "Enzymatic Approaches to Production of Sucrose from Starch," *Biotechnol. Bioeng.* 22:1504-1507 (1980).
33. Koaze, Y., Y. Nakajuma and T. Eida. "Improvement of Soybean Products of Microbial Means," in *Proc. Int. Symp. Conversion and Manufacture of Foodstuffs by Microorganisms*, (Tokyo, Japan: Saikon Publishing Co. Ltd., 1971), pp. 41-51.

34. Kargi, F., and M. L. Shuler. "An Evaluation of Various Flocculants for the Recovery of Biomass Grown on Poultry Waste," *Agriculture Wastes* 2:1-12 (1980).

35. Sinskey, A. J., and S. R. Tannebaum. ' Removal of Nucleic Acids in SCP," in *Single Cell Protein II* (Cambridge, MA: The M.I.T. Press, 1975), p. 158.

36. Romantschuk, H. "The PEKILO Process: Protein for Spent Sulfite Liquor," in *Single-Cell Protein II* (Cambridge, MA: The M.I.T Press, 1975), pp. 344-356.

37. Harrison, D. E. F., T. G. Wilkinson, S. J. Wrens and J. M. Harwood. "Mixed Bacterial Cultures as a Basis for Continuous Production of SCP from C_1 Compounds," in *Continuous Culture 6. Applications and New Fields*, A. C. R. Dean, D. C. Ellwood, C. G. T. Evans and J. Melling, Eds. (New York: Ellis and Horwood Publishing, 1976).

38. Riviere, J. "Microbial Proteins" *Industrial Applications of Microbiology*, M. O. Moss and J. E. Smith, Translators/Eds. (New York: John Wiley & Sons, Inc., 1977), pp. 105-149.

39. Meyer, O. "Using Carbon Monoxide to Produce Single-Cell Protein," *Bioscience* 30(6):405-407 (1980).

40. Bernstein, S., and C. M. Tzeng. "Commerical Production of Protein by the Fermentation of Acid and/or Sweet Whey," U.S. Environmental Protection Agency, Industrial Environmental Research Laboratory, EPA-600/2-77-133, Cincinnati, OH (July 1977).

41. Shipman, R. H., L. T. Fan and I. C. Kao. "Single-Cell Protein Production by Photosynthetic Bacteria," in *Advances in Applied Microbiology*, Vol 21, D. Perham, Ed. (New York. Academic Press, Inc., 1977) pp. 161-183.

42. Forney, L. J., and C. A. Reddy. "Fermentative Conversion of Potato Processing Wastes into a Crude Protein Feed Supplement by Lactobacilli," in *Developments in Industrial Microbiology*, Vol. 18, L. A. Underkofler and M. L. Walf, Eds. (Washington, DC. American Institute of Biological Sciences, 1977), pp. 135-143.

43. "LSU Promotes Protein-From-Cellulose Process," Tech. Note, *Chem. Eng. News* 20 (February 18, 1974).

44. Srinivasan, V., Louisiana State University, Baton Rouge, LA. Personal Communication (October 1980).

45. Spano, L. A. "Enzymatic Hydrolysis of Cellulosic Wastes to Fermentable Sugars and the Production of Alcohol," *J. Coating Technol.* 50(637): 71-78 (1978).

46. Coppinger, E., J. Brautigam, J. Lenart and E. D. Baylon. "Report on the Design and Operation of a Full-Scale Anaerobic Dairy Manure Digester," SERI/FR-312-471 prepared for DOE by Ecotope Group, Seattle, WA (1979).

47. Engel, A. J., Pennsylvania State University, PA. Personal Communication (August 1980).

48. "Oil from Algae: Blue Sky or Realistic Goal?" *Chem. Wk.* (July 18, 1979), pp. 43-44.

49. Cysewski, G. R., and C. R. Wilke. "Rapid Ethanol Fermentations Using Vacuum and Cell Recycle," *Biotechnol. Bioeng.* 14:1125-1143 (1977).

50. Jewell, W. J. "Future Trends in Digestor Design," paper presented at the first International Symposium on Anaerobic Digestion, Cardiff, Wales, September, 1979.

51. Jewell, W. J. "Low Cost Methane Generation of Small Farms." paper presented at the 3rd Annual Symposium on Biomass Energy Systems, June, 1979.
52. Jewell, W. J., S. Dell'Orto, K. J. Fanfoni, T. D. Hayes, A. P. Leuschner and D. F. Sherman. "Anaerobic Fermentation of Agricultural Residues: Potential for Improvement and Implementation—Final Report," Vol II, Project No. DE-ACO2-76ET20051 (1980).
53. Fraser, M. D. "The Economics of SNG Production by Anaerobic Digestion of Specially Grown Plant Matter," in *Symposium Papers: Clean Fuels from Biomass on Wastes*, sponsored by the Institute of Gas Technology, Washington, DC (1977), pp. 425-439.
54. Yen, T. F. "Microbial Oil Shale Extraction," in *Microbial Energy Conversion*, Schlegel and Baruea, Eds. (Elmsford, NY: Pergamon Press, Inc., 1977).
55. Yen, T. F., University of Southern California, Los Angeles, CA., Personal Communication (October 1980).
56. Pfeffer, J. T. "Methane from Urban Wastes—Process Requirements," in *Microbial Energy Conversion*, Schlegel and Baruea, Eds. (Elmsford, NY: Pergamon Press, Inc., 1977), pp. 139-155.
57. Sutherland, I. W., and D. C. Ellwood. "Microbial Exopolysaccharide Industrial Polymers of Current and Future Potential," in *Symp. Soc. for General Microbiogy*, A. T. Bull, D. C. Ellwood and C. Ratledge, Eds., No. 29, Microbial Technology: Current State, Future Prospects (New York: Cambridge University Press, 1979), pp. 107-150.
58. Silman, R. W., and P. Rogovin. "Continuous Fermentation to Produce Xanthan Biopolymers: Laboratory Investigation," *Biotechnol. Bioeng.* 12: 75-83 (1970).
59. Su, T. M. "Bioconversion of Cellulosic Fiber to Ethanol," in *Proc. 4th Joint US/USSR Conf. on the Microbial Enzyme Reactions*, US/USSR Joint Working Group on the Production of Substances by Microbial Means, NTIS Publication No. PB 132913 (1979), pp. 121-135.
60. Pye, R. E., B. Hagerdal and J. Ferchak. "High-Temperature Saccharification of Cellulose by *Thermoactinomyces* Cellulase for Liquid Fuel Production," in *Proc. 4th Joint US/USSR Conf. on the Microbial Enzyme Reactions*, NTIS Pub. No. PB 132913 (1979), pp. 383-400.
61. Van Den Berg, L., and C. P. Lentz. "Methane Production During Treatment of Food Plant Wastes by Anaerobic Digestion," paper presented at the 9th Annual Waste Management Conference, Syracuse, NY, 1977.
62. Van Den Berg, L., and C. P. Lentz. "Food Processing Waste Treatment by Anaerobic Digestion," National Research Council of Canada Publication No. 15981 (1977).
63. Ghosh, S., and D. L. Klass. "Two-Phase Anaerobic Digestion," in *Symposium Papers: Clean Fuels from Biomass and Wastes*, sponsored by Institute of Gas Technology, Washington, DC (1977), pp. 373-415d.
64. Bennett, M. A., and M. H. Weetall. "Production of Hydrogen Using Immobilized *Rhodospirillium rubrum. J. Solid Phase Biochem.* 1(2): 137-142 (1976).
65. Benemann, J. R., K. Miyamoto and P. C. Hallenbeck. "Bioengineering Aspects of Biophotolysis," *Enzyme Microbiol. Technol.* 2: 103-111 (1980).

66. Miyamoto, K., P. C. Hallenbeck and J. R. Benemann. "Solar Energy Conversion by Nitrogen-limited Cultures of Anabaena cylindrica," *J. Fermentation Technol.* 57(4):287-293 (1979).

67. Zürrer, H., and R. Bachofen. "Hydrogen Production by the Photosynthetic Bacterium *Rhodospirillum rubrum.*," *Appl. Environ. Microbiol.* 37(5):789-793 (1979).

68. Jeris, J. S., and R. W. Owens. "Pilot-Scale, High-Rate Biological Denitrification," *J. Water Poll. Control Fed.* 47(8):2043-2057 (1975).

69. Tracy, K. D., and T. G. Zitrides. "Mutant Bacteria Aids Exxon Waste System," *Hydrocarbon Proc.* (October 1979).

70. "Tall Order in Waste Treatment," *Chem Wk.* 67-68 (November 10, 1976).

71. Baig, N., and E. M. Grenning. "The Use of Bacteria to Reduce Clogging of Sewer Lines by Grease in Municipal Sewage," in *Biological Control of Water Pollution*, Tourbeir and Pierson, Eds. (Philadelphia: University of Pennsylvania Press, 1976), pp. 245-252.

72. Weissman, J. C., D. M. Eisenberg and J. R. Benemann. "Cultivation of Nitrogen-Fixing Blue-Green Algae on Ammonia-Depleted Effluents from Sewage Oxidation Ponds," *Biotechnol. Bioeng. Symp. No. 8.* 299-316 (1978).

73. Tuovinen, O. H., The Ohio State University, Columbus, OH. Personal Communication.

74. Whittington, D., and W. R. Taylor. "Regulation and Restoration of In-Situ Uranium Mining in Texas," in *Proc. South Texas Uranium Seminar*, Am. Inst. of Mining, Metallurgical and Petroleum Engineers (1979), pp. 7-10.

75. Gale, N. L., and B. G. Wixson. "Wastewater Discharge Sites from Mining Operations in the "New Lead Belt" of Missouri," Society of Mining Engineers of AIME, Preprint No. 77-AG-350 (1977).

76. Miles Product Literature on TAKA-THERM and DIAZYME L-100.

77. Freeman, J., Economy Alcohol Fuel Supplies, Dayton, OH. Personal Communication.

78. Friend, B. A., and K. M. Shahani. "Whey Fermentation." *New Zealand J. Dairy Sci. Technol.* 14:143-152 (1979).

79. Wilke, C. R., G. R. Cysewski, R. D. Young and U. Von Stockar. "Utilization of Cellulosic Materials through Enzymatic Hydrolysis. II. Preliminary Assessment of an Integrated Processing Scheme," *Biotechnol. Bioeng.* 18:1315-1323 (1976).

80. Lipinsky, E. S., and J. H. Kitchfield. "Single-Cell Protein in Perspective," *Food Technol.* 28:16-20 (1974).

81. Weetall, H. H. "Immobilized Enzymes and Their Application in the Food and Beverage Industry," *Process Biochem.* 10:3-8 (1975).

82. Davis, E. N., and L. L. Wallen. "Viscous Products from Activated Sludge by Methanol Fermentation," *Appl. Environ. Microbiol.* 32:303-305 (1976).

83. Roth, W. B. "Methanol Treated Activated Sludge as an Agricultural Chemical Carrier," U.S. Patent No. 4,065,287 (December 27, 1977).

84. Griffith, W. L., and A. L. Compere. "Continuous Lactic Acid Production Using a Fixed-Film System," *Develop. Ind. Microbiol.* 18:723-726 (1977).

85. Cooney, C. L., D. I. C. Wang, S. Wang. J. Gordon and M. Jimminez. "Simultaneous Cellulose Hydrolysis and Ethanol Production by a Cellulolytic Anaerobic Bacterium," *Biotechnol Bioeng. Symp. No. 8* 103-114 (1978).
86. Kelly, D. P. "Extraction of Metals from Ores by Bacterial Leaching: Present Status and Future Prospects," in *Microbial Energy Conversion*. Schlegel and Barnea, Eds. (Elmsford, NY: Pergamon Press, Inc., 1977), pp. 329-338.
87. Brierley, C. L. "Thermophilic Microorganisms in Extraction of Metals from Ores," in *Developments in Industrial Microbiology*, Vol. 18 (Washington, DC: American Institute of Biological Sciences, 1977), pp. 273-284.
88. Brierley, C. L., J. A. Brierley and L. E. Murr. "Using the SEMI in Mining Research," *R/D* 24(8):24-28 (1979).
89. Cameron, R. D., and F. A. Koch. "Trace Metals and Anaerobic Digestion of Leachate," *J. Water Poll. Control Fed.* 52(2):282-292 (1980).
90. Guilbault, C. G., S. S. Kuan, W. Fung, B. Chen, A. Kalmar and H. Winartpurtra. "Bioprobes," in *Proc. 4th Joint US/USSR Conf. on the Microbial Enzyme Reactions, US/USSR Joint Working Group on the Production of Substances by Microbiological Means*, NTIS Publ. No. PB80-132913 (1979).
91. Rechnitz, G. A., T. L. Riechel, R. K. Kobos and M. E. Meyerhoff. "Glutamine-Selective Membrane Electrode that Uses Living Bacterial Cells," *Science* 199:440 (1978).
92. Karube, I. T., Martsunga, S. Mitsuda and S. Suzuki. "Microbial Electrode BOD Sensor," *Biotechnol. Bioeng.* 19:1535-1547 (1977).